LYSENKO'S GHOST

LYSENKO'S GHOST

EPIGENETICS AND RUSSIA

Loren Graham

Harvard University Press

Cambridge, Massachusetts

London, England

2016

First printing

Library of Congress Cataloging-in-Publication Data

Names: Graham, Loren R., author.
Title: Lysenko's ghost : epigenetics and Russia / Loren Graham.
Description: Cambridge, Massachusetts : Harvard University Press, 2016. |
    Includes bibliographical references and index.
Identifiers: LCCN 2015038198 | ISBN 9780674089051 (alk. paper : alk. paper)
Subjects: LCSH: Lysenko, Trofim, 1898–1976. | Epigenetics—Russia (Federation)
Classification: LCC QH438.5 .G73 2016 | DDC 576.5/3—dc23
LC record available at http://lccn.loc.gov/2015038198

To Meg and Kurt

Lysenko was right. The problem is that he was a criminal who was responsible for the suffering and deaths of his colleagues.

—v. s. BARANOV, Head, Laboratory, Academy of
Medical Sciences, St. Petersburg, Russia,
author interview, June 20, 2014

CONTENTS

LYSENKO'S GHOST

INTRODUCTION

RECENT EVENTS in the science of heredity have reignited controversies in Russia and elsewhere about the scientific significance of the Russian agronomist Trofim Lysenko (1898–1976). Lysenko, one of the most infamous scientists of the twentieth century, believed in the inheritance of acquired characteristics (that is, that traits acquired during the lifetime of an organism can be passed on to its offspring) and attempted to apply that doctrine to the cultivation and raising of crops and livestock. He exercised increasing influence in Russian biology from the mid-1930s onward, and in 1948 he was enthroned as ruler of the field, with the approval of Joseph Stalin himself.

But his fall was as dramatic as his rise. In 1965, Russian geneticists declared him a fraud and condemned him for causing great damage to Soviet agriculture. Furthermore, these scientists accused him of being responsible for the imprisonment and deaths of many of the best geneticists in the Soviet Union. "Lysenkoism" became a synonym for pseudoscience in Russia and indeed around the world.[1]

Lysenko's identity as a fraud remained unchallenged for decades, until just around the turn of the twenty-first century. Then, advances in the new science of epigenetics led to a reassessment of his legacy. Research in epigenetics has called into question the dogma, established with the rise of genetics in the early twentieth century, that hereditary information is contained almost exclusively within the DNA sequence of genes and that events experienced during an organism's lifetime cannot be inherited unless they alter this sequence. Scientists discovered, however, that modifications to the DNA molecule that do not affect its sequence (such as the attachment of certain chemical groups) can profoundly affect gene activity, for example, by turning genes "on" and "off." These modifications can be environmentally induced, and in some instances, they can be passed on to subsequent generations. So it appears that acquired characteristics may indeed be heritable after all.

These new developments in the science of heredity are now being debated worldwide. Some enthusiastic scientists have spoken of a "revolution" in our concepts of heredity,[2] but more sober critics have pointed out that much is still not known about how pervasive and important epigenetic phenomena are in inheritance, or how they work.[3] Epigenetics is a controversial subject, with much disagreement among experts. The nature of the debate in Russia is different than it is elsewhere: because of the history of the Lysenko affair, the discussions are much more sharp-edged and political. Articles in recent Russian publications have been headlined "Lysenko Was Right!" and "Lysenko's Views Confirmed by New Science!" The political component is clearly illustrated by some of the authors' identification as Stalinists and as people nostalgic for the old Soviet Union.

This book explores these recent events by focusing on three themes: the doctrine of the inheritance of acquired characteristics and its relationship to political ideas; the new field of epigenetics and how it is changing our views on the inheritance of acquired characteristics; and the impact of epigenetics in Russia, which has a particular history in relation to the doctrine of the inheritance of acquired characteristics. Some people may question whether these three themes fit together felicitously in one book. In my opinion, the three are so intertwined, and each so impacts the other, that it is necessary to include them all.

This book does not tell the history of the rise and fall of Lysenko. That story has been well described elsewhere.[4] It is, instead, an effort to describe changing views on inheritance in the West and attitudes

Trofim Lysenko. (Novosti/Science Source)

toward epigenetics and Lysenko in Russia in recent years. And it is also an effort to directly address the question, In light of the development of epigenetics, was Lysenko right after all? I attempt to answer this question on the basis of my study of Lysenko's views and our current knowledge of genetics and epigenetics. I have worked on the Lysenko phenomenon, off and on, for over fifty years and am one of the few people in the West still alive today who met Lysenko and talked to him about his work.

Writing this book has been an unusual, illuminating, and disconcerting experience for me as a historian of science. In my previous works, my concept of the proper role of the historian of science has usually been the same as that of most of my colleagues in the discipline. A common view among them is that it is not the task of historians of science to evaluate science; that should be left to the scientists. The proper role of the historian of science, many assert, is to portray and analyze the social context in which science develops—to illustrate how the views being expressed by scientists can be connected to the social, political, and ideological environment in which they lived and worked. Historians of science often shy away from the concept of "truth," knowing that we never possess it in an absolute sense and realizing that whatever we might call "true" at one point will likely be superseded by a different definition of "true" later.

Furthermore, it is obvious that the terms "true" and "Lysenko" are thick and multidimensional, meaning different things to different historical actors at different times. To some people, "true" has a specific epistemological meaning, while to others it is only a relative term. To some people, "Lysenko" means a particular genetic theory, the one he espoused, while to others the name means political interference in science, and not much else. These complications may suggest to some that my very posing of a question about the truth of Lysenko's claims, "Was Lysenko right after all?" is a mistaken quest.

It would have been easy, perhaps too easy, for me to write a history of recent attitudes toward genetics in Russia within the standard framework of a historian of science—to describe how Lysenko's views during Soviet times developed in the context of agricultural crisis, ideological militancy, political pressure, centralized control, and incomplete genetic knowledge; then I would trace the rise in Putin's Russia of a new version of Lysenkoism in a different social context, conditioned by growing nostalgia for Soviet times, a new set of political and nationalistic pressures, and increasing anti-Westernism. I would not pose the question of whether the new field of epigenetics reveals that Lysenko had some grasp on the truth. I would not judge, merely explain.

Such a treatment would hopefully be enlightening, but would it be adequate or sufficient? If I limited my treatment in this way, I would be approaching a complete relativism, showing that different contexts produce different results—which is all that the historian of science needs to demonstrate. After much thought and considerable agonizing, I found that I could not live with that conclusion. The question of whether Lysenko was "right" after all cannot, it seems to me, be dodged. It begs, even demands, to be addressed. I decided to directly confront the tension between my role as a historian of science and my desire to evaluate recent claims about Lysenko.

This book is an illustration of my belief that this tension between analysis and evaluation can, in some instances at least, be productive rather than destructive. We can never grasp absolute truth, but we can certainly grasp more or less substantiated views of nature. At any point in time we have the possibility, indeed, the obligation, to try to distinguish the more justifiable view of nature from the less justifiable one, knowing well that such distinctions may later be revised. I have tried, therefore, to perform a dual task: to show how the new Lysenkoism in Russia is a product of a particular post-Soviet context and also to evaluate its validity. I think it is good for each generation, at each moment in the development of science, to try to draw a bottom line and lay out the truth as that generation sees things at that moment. The following generation will tell us that the bottom line is somewhere else. That is the way we learn.

1 | THE FRIENDLY
SIBERIAN FOXES

This is a new and controversial direction of thought—here we could fall back into Lysenkoism. . . . The most dangerous possibility is that this direction might undermine genetics itself. —V. BARTEL'

REMOVING THE PADDED GLOVE protecting my hand from the cold, I reached out to the Siberian fox coming toward me. The fox licked my fingers and lowered itself as if wanting to be petted. I stroked the fox's sleek back. It responded as would a family dog, moving closer and shifting slightly to receive my full attention. It obviously enjoyed the contact and busily licked my other hand, undeterred by the glove still encasing it. Foxes are normally hostile to humans; this fellow seemed to love them.

The explanation of the almost sixty-year-old experiment that led to this result leads one into questions of science and politics extending from the 1940s to the present day. The controversies surrounding the experiment have never left—just taken on a new intensified form that would have dismayed its initiators. What they hoped to prove they are in danger of disproving, shocking not only themselves but the entire world of biology.

The man who started this experiment was Dmitri Belyaev, a Russian biologist caught in one of the most infamous controversies in science of the past century. Russian biology at the time of Belyaev's higher education in the 1940s was beset by a great controversy over genetics. A poorly educated and dogmatic agronomist named Trofim Lysenko was challenging genetics as it was known all over the world. He denied that genes were the main carriers of heredity, dismissed molecular biology, and denounced the founders of modern genetics, such as the American Thomas Hunt Morgan, as "bourgeois falsifiers." He preached a doctrine based on the inheritance of acquired characteristics, the idea that the traits an organism acquires during its lifetime can be passed on to its offspring. His views in 1948 received the imprimatur of Stalin himself and, as a result, the secret police repressed geneticists throughout the Soviet Union. Some were executed, while many others were imprisoned in labor camps, where a significant number died. Lysenko's most well-known opponent, Nikolai Vavilov, perished of starvation in such a place. Political power had overcome science, almost extinguishing classical genetics in the Soviet Union. Lysenko had triumphed.

Belyaev knew his genetics and firmly believed that Lysenko's views were false. Working as a young man in the Central Research Laboratory of Fur Breeding in Moscow, he rashly expressed his opposition. He was fired from his job as a result, but perhaps because of his junior status he was not arrested. Belyaev realized that he could not openly practice and research genetics in the Soviet Union at that time. He looked for a way to continue his research without thoroughly compromising his principles. In the 1950s the Soviet government created a new science city in faraway Siberia, near Novosibirsk. The political rulers in Moscow were over two thousand miles away. Attracted not only by the modern facilities but also by the relative freedom the remote location afforded, many talented scientists moved there.[1] They offered quiet shelter to a few geneticists fleeing Lysenkoism. Belyaev was one of them. Other anti-Lysenko biologists also gathered in Novosibirsk.[2]

Belyaev searched for a topic that was scientifically interesting but not politically dangerous. He felt drawn to the issue of animal domestication. He chose foxes, common in Siberia, and soon acquired several hundred, a population that formed the starting point for his experiments. By itself the domestication of animals did not raise red flags for Lysenko supporters. After all, Charles Darwin had written a book on the subject long before the very word "gene" was even invented. Lysenko sometimes called his approach "creative Darwinism."[3] Domestication of animals was based on artificial selection, a centuries-old practice in animal and plant breeding and one pursued by a Russian horticulturalist educated in the nineteenth century, Ivan Michurin, whom Lysenko much admired. In fact, Lysenko sometimes called his approach "Michurinist biology."[4] Surely, the selective breeding of animals was safe research.[5]

Safe, but not compromised. Belyaev believed that by researching the domestication of foxes he could stay loyal to genetics as it was practiced elsewhere instead of to Lysenko's idiosyncratic views. Belyaev, contrary to Lysenko, remained convinced of the existence and importance of genes; his fox domestication experiment was based on the principles of genes, mutation, recombination, and artificial selection. The inheritance of acquired characteristics that Lysenko emphasized had nothing to do with it. Belyaev did not dream that his experiment might eventually question some of his basic principles.

Belyaev divided the wild foxes he had collected into three groups: the fiercest ones, hostile to all human approach; foxes that did not bite or flee when fed by humans but obviously did not like them; and foxes that showed some interest in humans, if only by whining when they approached. Belyaev then submitted these foxes to strong artificial selection pressure, breeding the least hostile foxes with each other and then, in each generation, continuing to select for breeding those foxes that showed the least hostility and, eventually, the most friendliness. By the sixth generation, a new class of foxes had emerged that Belyaev christened the "domesticated elite." They licked and sniffed humans like dogs. By the tenth generation, 18 percent of the

Russian geneticist Dmitrii Belyaev with tame silver foxes. (RIA Novosti/
Science Source)

fox pups were members of this elite class. By the twentieth generation, 35 percent belonged to this class. The foxes in this group were
not just very friendly; many had changed physically and had floppy
ears and wagging tails. Today, after over forty generations, 70 to
80 percent are in this highest class. In fact, the majority of the foxes
became so friendly that Belyaev's research institute began selling some
as household pets, raising money for further experimentation.[6]

Belyaev explained these results quietly in terms of classical genetics. He theorized that he was selecting from a group of foxes that had genetic variation in temperament, progressively increasing the proportion of the friendliest. He assumed that particular genes, acting either alone or in groups, were responsible for this friendly nature.

I visited Belyaev and his fox farm several times. Belyaev was director of the Institute of Cytology and Genetics, which was located in the science city of Akademgorodok near Novosibirsk. The fox farm itself was out in the countryside, several miles from Belyaev's office. In 1976 I journeyed out to the farm without Belyaev. There I met the assistants who were in charge of the foxes' day-to-day care. I asked one, a kindly woman dressed in the country attire of *valenki* (felt boots), a heavy coat, and a head shawl, why she thought the foxes were so friendly. She replied, "Because we take such good care of them and because we love them. We constantly stroke them, supply them the best food, give all of them names, call them individually by these names, and show our affection for them. They respond by returning our love, and that love becomes hereditary."

This answer surprised me because it made no reference to genes and was clearly based on the inheritance of acquired characteristics—a doctrine their boss, Belyaev, denied. In fact, it was a view thoroughly in accord with Lysenko's. After all, Lysenko claimed that he could get cows and their progeny to give more milk by caring for them attentively, keeping their stalls clean, and feeding them copiously. Genetic pedigree meant nothing to Lysenko. It did not seem to mean much to Belyaev's assistants, either.

I returned to Belyaev's office and quizzed him on this issue. "Do you know," I asked, "that some of the members of your staff out at the fox farm are supporters of Lysenko? They believe that the foxes become friendly because of the care they give them, not as a result of genetic selection." Belyaev laughed and said yes, he knew. "But it

Author feeding a domesticated fox in Siberia. (Courtesy of the WGBH Media Library and Archives)

is not correct to call them followers of Lysenko," he continued. "They are simply supporters of the doctrine of the inheritance of acquired characteristics. It is a mistaken view, but does no harm. In fact, they probably take better care of the foxes than I would, since they attribute to that care a significance that I do not. So they make excellent lab assistants. I want attentive helpers who take good care of the foxes and laboratory facilities."

I would remember this conversation years later when the full significance of the new science of epigenetics dawned on me. According to epigeneticists, the effects of an organism's life experiences *can* in some instances become heritable, at least for a few generations. Contrary to Belyaev, it appears that the inheritance of acquired characteristics does indeed occur.

A well-known experiment in the West in the early study of epigenetics has similarities to Belyaev's Siberian fox experiments. Michael Meaney of McGill University has maintained that in certain rat litters, the pups that receive the most licking and grooming from their mothers become adult rats that in turn lavish attention on their own pups, and this continues in future generations.[7] Meaney proposed that this tendency was a trait connected with gene expression controlled by chemical groups attached to the rats' DNA.[8] And these "chemical attachments" resulted, he thought, from behavior—in this case, licking and grooming. Meaney pointed to the possibility of the inheritance of acquired characteristics, obtained in a way highly similar to what the caretakers of the Siberian foxes told me long ago. Instead of a scientific explanation, those caretakers merely said, "We love the foxes and they love us in turn, and this becomes hereditary."

Does epigenetics help explain the Siberian fox experiment? Could the care the infant foxes received from their caretakers influence the behavior of their progeny? Does the Siberian fox experiment cause us to rethink the question of Lysenko? We do not know complete answers to these questions yet. Much research needs to be done. But these issues are beginning to bother the people who continue the experiment in Siberia today, long after the death of Belyaev in 1985.

In 2007 the Institute of Cytology and Genetics celebrated the ninetieth anniversary of Belyaev's birth. Celebrating the birth anniversaries of leading scientists is standard practice in Russian science. At such occasions the speakers usually praise the deceased, giving almost scripted and predictably laudatory performances. But several of the speakers at the 2007 Novosibirsk conference referred to "epigenetic inheritance." Two non-Russian participants, Eva Jablonka and

Marion J. Lamb, were already well known for claiming that modern biology must make room for the doctrine of the inheritance of acquired characteristics.[9] Although this point of view would not disturb the caretaker of the foxes I talked to long ago, it definitely disturbed several of the Russian academically trained geneticists at the conference. After all, they and their teacher Belyaev had struggled bravely against Lysenko and had seen colleagues imprisoned as a result. One observed that the science of epigenetics was provoking new questions about the fox experiment and raising the possibility of "falling back into Lysenkoism":

> The most dangerous issue here is that these new developments may undermine genetics itself. Lysenko said that the conditions of the environment could change everything. Epigenetics approaches this view, maintaining that the change of an organism's characteristics occurs not under the influence of mutations and recombinations, but thanks to a change in the expression of genes.[10]

And he noted that the expression of genes can, according to the epigeneticists, be influenced by the environment.[11]

The science of heredity is under debate worldwide, especially in the wake of the new science of epigenetics. This scientist's reaction highlights a key question: If epigenetic research shows that acquired characteristics can be inherited, a doctrine that Lysenko promoted, does that mean Lysenko was right?

2 | THE INHERITANCE OF
ACQUIRED CHARACTERISTICS

> The concept of the inheritance of acquired characters is a notion that
> had been held almost universally for well over two thousand years.
>
> —CONWAY ZIRKLE

THE DOCTRINE of the inheritance of acquired characteristics has a
long history, being associated in its later years mainly with French
naturalist Jean-Baptiste Lamarck. After the rise of genetics, however,
it fell out of favor, especially in the West. For most of the twentieth
century, the concept was in disrepute in establishment biology. Most
leading biologists of the time, embracing the modern synthesis of
Darwinism and Mendelism that occurred in the first part of the
century, derided the inheritance of acquired characteristics and ridiculed those who spoke positively of it. They believed they could
explain inheritance on the basis of natural selection, mutation, and
recombination. The scandals of the Paul Kammerer affair (1926) and
the Lysenko affair (roughly 1936–1965), to be discussed in Chapters 3
and 6, respectively, painted the inheritance of acquired characteristics in the worst possible colors. Its reputation was that it was both
false and reprehensible, supported mainly for left-wing political reasons associated with Kammerer and Lysenko. Because it was discredited in the West, it became associated with discredited scientists and
disassociated from respected scientists. A scientist risked his or her
reputation by seriously investigating the inheritance of acquired characteristics. Cold War politics played a role in all this.[1]

A survey of thirty of the most widely used college textbooks of genetics between 1962 and 1990 found none that cited examples of the inheritance of acquired characteristics.[2] The following was a typical treatment of the subject: "Lamarck's hypothesis of the inheritance of acquired modifications has been discarded because no molecular mechanism exists or can be imagined that would make such inheritance possible."[3]

And yet this disreputable concept has made a comeback in the twenty-first century. What is the inheritance of acquired characteristics? A classical definition states that "modifications acquired by an organism can be inherited by offspring."[4] We now know that according to epigenetics such features can, at least in some instances, be passed on. However, some biologists still resist. "Epigenetic inheritance is not really the same as the inheritance of acquired characteristics," they say, "since the underlying DNA is not changed, just the expression of certain genes." They point out that epigenetic marks are often erased at reproduction, and if they are not, no one knows yet how many generations will bear the marks. Furthermore, epigenetic inheritance is more evident in simple organisms than in humans and other mammals. All this is true, but since what happens fits beautifully with the classical definition of "features acquired during an organism's lifetime are passed on," a forthright, unprejudiced reaction should be, "Yes, in these instances, which are still not sufficiently explored, the inheritance of acquired characteristics does occur." Biologists are often still reluctant to use the term "inheritance of acquired characteristics" because of its unsavory reputation. Instead, they speak of "epigenetic inheritance."

In 2013 I attended a lecture at Harvard Medical School at which Scott Kennedy of the University of Wisconsin asserted, "Many examples of the transgenerational transfer of epigenetic information have now been documented."[5] Kennedy described his work with the worm *C. elegans* and how such inheritance had persisted for six to eight generations. He said he thought the length of time could be extended. Never once in his lecture did he use the words "inheritance of acquired characteristics." He simply described such inheritance without naming it.

As time goes on, the twentieth-century denial of the inheritance of acquired characteristics is likely to be considered an odd detour in biological thought. Belief in the doctrine goes back to the classical period, to Hippocrates and beyond. One scholar who studied the history of the concept of the inheritance of acquired characteristics in detail observed it to be "a notion which had been held almost universally for well over two thousand years."[6] Among the dozens of supporters of the doctrine, he included Hippocrates, Pliny, Aristotle, Roger Bacon, Andreas Vesalius, John Rey, William Godwin, Charles Lyell, and Charles Darwin himself. Of course, these thinkers had very different concepts of what the inheritance of acquired characteristics actually meant and the means by which it worked. Darwin, for example, reluctantly accepted it as a way of explaining certain anomalies in his evolutionary description, emphasizing that in his opinion natural selection was the "far greater power"; he even devised a scheme, which he called "pangenesis," to describe how it worked. And in successive editions of *The Origin of Species,* he gradually increased his emphasis on the inheritance of acquired characteristics.

Against this background, it is striking that today when one mentions the inheritance of acquired characteristics to biologists, they almost invariably answer, "Oh, you mean Lamarckism." Why should the name of Lamarck, a French biologist of the late eighteenth and early nineteenth centuries, be almost synonymous with a concept that has been supported by scientists and thinkers for millennia?

JEAN-BAPTISTE LAMARCK

Jean-Baptiste Lamarck (1744–1829) was a great scientist—but not for the reasons that people usually remember today.[7] His scheme of taxonomy, classifying invertebrates such as worms, spiders, and mollusks, was groundbreaking. As a botanist his method of analysis was superior to that of Linnaeus, and his three-volume work on the flora of France was both popular and scientifically significant. Although originally a believer in the fixity of species, he gradually moved toward an evolutionary view, at the same time displaying a speculative element in his thinking. The idea of "increasing complexity" in nature led him to seek a mechanism for change. He rejected vitalism or theism as this mechanism and sought a naturalistic explanation. In 1809, as his health began to fail at the age of sixty-five, he published his *Philosophie zoologique,* the work that is best known today and in which can be found a full theory of evolution. Many later discussions of Lamarck have concentrated on this work, ignoring his earlier multivolume studies of classification. Lamarck amplified his views on evolution in the introduction to his seven-volume 1815 work, *Histoire naturelle des animaux sans vertèbres.* In this introduction he discussed more fully the element of his scheme that has drawn the most attention; namely, the inheritance of acquired characteristics. The ahistorical nature of this attention can be seen by noting that almost all of his contemporaries also believed in such inheritance. Nonetheless, Lamarck has been so identified with this principle that "Lamarckism" conceived in this way has become a standard part of the biological vocabulary. That, unfortunately, cannot be changed. Such is the weight of usage over accuracy.

Lamarck's evolutionary scheme was based on two factors: a natural tendency toward complexity and the influence of the environment.[8] His discussion of why giraffes have long necks is most often cited and is frequently contrasted with Darwin's theory. Lamarck

French botanist Jean-Baptiste Lamarck was an early nineteenth-century proponent of the inheritance of acquired characteristics.
(Science Source)

believed that the lengthening of giraffes' necks when they stretched to seek food in high tree branches could be inherited, resulting in longer necks in giraffes in a cumulative process. Darwin, on the other hand, said that in any population of giraffes there are, as a result of random variation, some with longer necks than others; through natural selection those with long necks survived more often than those with short ones, resulting in longer necks among giraffes over the succeeding generations. The important distinction here is that Lamarck thought that features acquired during the life of an individual giraffe could be evolutionarily important, while Darwin emphasized selection out of random natural variation. This example is somewhat simplistic since Darwin never entirely denied the inheritance of acquired characteristics, but the explanatory power of the example is so great that it is a standard part of many textbooks of biology. These textbooks further note that advances in genetics by scientists such as August Weismann have shown that the inheritance of acquired characteristics is an impossibility.

AUGUST WEISMANN AND GERM PLASM THEORY

One of the most significant turning points in late nineteenth-century biology came with German evolutionary biologist August Weismann (1834–1914), who portrayed a sharp distinction, even a barrier, between the germ cells and the rest of an organism's body.[9] As was the case with Lamarck, so it has been with Weismann: what the man actually said (wrote) and what the world has taken him for having said are slightly different. Weismann has been almost universally described as having developed a view—the "germ plasm theory"—in which inheritance is transmitted only through the germ cells. Other cells of the body, the somatic cells, do not play a role in heredity. This hereditary process proceeds in only one direction: germ cells produce somatic cells and are not influenced by anything that happens to the somatic cells. This concept would rule out the inheritance of acquired characteristics. And although Weismann obviously knew nothing about DNA, the concept attributed to him seems to prefigure the "central dogma" of molecular biology: hereditary information cannot be transferred back from protein to nucleic acid.

Weismann is also noted for his alleged "disproof" of the inheritance of acquired characteristics. He performed a famous experiment in which he cut off the tails of hundreds of mice for twenty-two generations. Yet the mice in the last generation had tails just as long as those in the first. Although one could question whether mutilation is a good stand-in for environmental influences, Weismann's "mouse experiment" has been cited endlessly as a refutation of the inheritance of acquired characteristics.

A careful study of Weismann's work reveals that the simple characterization given above is inaccurate. One historian of science, R. G. Winther, even said that "Weismann was not a Weismannian." It turns out that Weismann did accept the influence of the environment on hereditary material. Nonetheless, just as in the case of Lamarck, the power of usage has overwhelmed accuracy. Winther believed that later biologists "reinterpreted Weismann in a manner suitable to their purposes."[10] Thus, Weismann is known in biology as the man who helped discredit Lamarckism. Nonetheless, partisans of Lamarckism continued to promote their views well into the twentieth century. The next discrediting would come with Paul Kammerer in 1926.

Despite its inaccuracies, this "Weismannian" understanding of heredity helped clear the underbrush of often romantic and idealistic versions of the inheritance of acquired characteristics and laid the groundwork for molecular biology. Only after this new foundation had been securely laid would it become apparent that "Weismannism" contained what epigeneticists would see as gross exaggerations.

German evolutionary biologist August Weismann advanced the germ
plasm theory, which contradicted Lamarck's view of genetics.
(Science Source)

Under assault from Weismannism, Lamarckism in Western Europe in the late nineteenth century was increasingly on the defensive. Weismann had a running debate on the subject with the popularizer Herbert Spencer, who eloquently defended the inheritance of acquired characteristics.[11] Spencer attracted the interest of the educated reading public, but he was not a research scientist, and several influential biologists found Weismann's arguments more persuasive. In 1909 the Danish biologist Wilhelm Johannsen introduced the word "gene" and the concepts of "genotype" (the genetic makeup of a cell) and "phenotype" (the observable characteristics of an organism that result from its genotype and the environment). During the next twenty years, the science of genetics developed rapidly, especially in the hands of Thomas Hunt Morgan at Columbia University and William Bateson at Cambridge University. Both vocally opposed the inheritance of acquired characteristics.

THE INHERITANCE OF ACQUIRED CHARACTERISTICS IN RUSSIA

The discussion of the inheritance of acquired characteristics was an old subject in Russia that went back long before the Russian Revolution of 1917. Just as was the case in Western Europe and America in the nineteenth century, many Russian naturalists believed strongly in the doctrine. An example was K. F. Rul'e, a zoologist and professor at Moscow University, who wrote in 1850, "The influence of the external world on the animal kingdom is strengthened in progeny, becomes hereditary. . . . The most trainable animals are those whose parents were trained."[12] The views of the popular German biologist and philosopher Ernst Haeckel, a devout Lamarckist, were well known to educated Russians in the late nineteenth century.

Many Russians did not see a conflict between Lamarckism and Darwinism; both were evolutionary theories, and to those who saw

American biologist Thomas Hunt Morgan won the Nobel Prize in 1933 for his work on the role of the chromosome in heredity. (Science Source)

"evolution" as the radical and novel idea of the times, that was the most important thing. Frequently, Russian biologists who considered themselves dedicated Darwinists included within their views some role for the inheritance of acquired characteristics, just as Darwin

English geneticist William Bateson. (SPL/Science Source)

German biologist and naturalist Ernst Haeckel. (Wellcome Images/Science Source)

himself did. Many agricultural scientists in Russia thought they saw evidence of the inheritance of acquired characteristics in their work. Often cited was the effect of milking on dairy cows; it was thought that daily milking led hereditarily to larger udders and greater milk capacity.

In Russia the doctrine of the inheritance of acquired characteristics did not suffer the enormous setback at the end of the nineteenth and the beginning of the twentieth century that it did in Great Britain and the United States. Educated Russians of course learned about the controversies over the issue in the West but often through edited and translated presentations that favored Lamarckism. The great debate between Weismann and Spencer in 1893 in the English journal *Contemporary Review* was supposedly reproduced in the Russian journal *Nauchnoe obozrenie* (Science Review); however, the editor of that journal, M. M. Fillippov, a strong supporter of Lamarckism, gave the advantage in the debate to Spencer. Fillippov published Spencer's criticism of Weismann in full, but he deleted two-thirds of Weismann's reply. Furthermore, Fillippov made no editorial comments about Spencer's presentation, but he appended critical comments to Weismann's rejoinder, such as "Weissman cited no convincing facts" and "His final deduction was completely unfounded."[13]

Part of the reason that Lamarckism still exerted such influence on Russian biological thought in the early twentieth century was the slowness with which the new genetics developed by Morgan and Bateson in the United States and England penetrated Russia. Both Morgan and Bateson vehemently opposed the inheritance of acquired characteristics and vowed to replace it with a new rigorous genetics based on germ cells and chromosome studies. Basically, there was no community of "geneticists" in Russia until about the time of the Russian Revolution in 1917, and even after that, the geneticists formed a small and embattled community.

British philosopher and sociologist Herbert Spencer. (Science Source)

Russia's most famous scientist in the 1920s, the Nobel laureate physiologist Ivan Pavlov, was a firm believer in the inheritance of acquired characteristics.[14] Trained in the spirit of nineteenth-century physiology, a field in which most specialists supported some form of Lamarckism, Pavlov paid little attention to the rapid progress being made in the early twentieth century in genetics and the increasing skepticism of most geneticists toward the inheritance of acquired characteristics. He saw genetics as a "foreign, borderland sphere of biology."[15] He was busy with pathbreaking research on animal behavior and the difference between "conditional" (a learned response to a specific stimulus) and "unconditional" (an instinctive response to a specific stimulus not dependent on previous learning) reflexes.[16] He even believed that over several generations conditional reflexes could become unconditional and therefore a part of the innate heredity of an organism. This belief led him to one of the most embarrassing episodes in his career.

One of Pavlov's colleagues in his laboratory, Nikolai Studentsov, did work on the inheritance of acquired characteristics in 1923 and 1924 that at first Pavlov found very exciting. In order to study changes over a number of generations of animals, Studentsov worked with mice rather than dogs, Pavlov's favorite animal. Mice breed much more rapidly than dogs.

Studentsov studied the number of repetitions required by successive generations of mice to develop a conditional reflex. He placed mice in cages and always sounded a buzzer before feeding them. Studentsov wanted to answer the question: How many repetitions of this double operation, the buzzer always sounding before the food appears, are necessary before the mice, hearing the buzzer, run to the feeding spot before the feed bag is inserted? Studentsov maintained that successive generations of mice required fewer and fewer repetitions. According to him, in the first generation of mice the

Russian biologist Ivan Pavlov. (Science Source)

buzzer had to be sounded, on average, 298 times before the mice began to anticipate the food automatically; in the second generation, 114 repetitions were required; in the third generation, 29; in the fifth generation, only 6. Therefore, said Studentsov, by means of the inheritance of acquired characteristics, a conditional reflex ("food follows buzzer") was converted into an inherited unconditional reflex that was passed on to later generations of mice.

In 1923 Pavlov toured Western Europe and the United States and gave many lectures, both academic and popular. Frequently, he cited Studentsov's research as evidence for the validity of the inheritance of acquired characteristics. Pavlov stated in these lectures, "I think it very probable that after some time a new generation of mice will run to the feeding place on hearing the buzzer with no previous lesson."[17]

One of the places where Pavlov expressed such views was in Woods Hole, Massachusetts, where many prominent biologists gathered in the summer to discuss research. Among Pavlov's listeners was Thomas Hunt Morgan. Historians believe that Morgan and Pavlov differed in the discussion after the lecture, but there is no record of what was said. What is clear is that Morgan was upset and that his displeasure increased when Pavlov went on to a physiology congress in Edinburgh and said the same thing. Morgan joined with William Bateson in questioning Studentsov's work and Pavlov's interpretation of it. They pointed out that Studentsov did not follow rigorous methodology, for example, not using a control group of untrained mice, and therefore the results were not trustworthy. An obvious counterhypothesis to Pavlov's was that the learning going on in these experiments was learning by Studentsov, not by his mice. In other words, Studentsov was becoming increasingly skilled at training mice. As a result, the number of repetitions necessary for the conditional reflex to be established was diminishing.

Morgan even made fun of Pavlov. In an article titled "Are Acquired Characters Inherited?," he stated rhetorically:

> How simple would our educational questions become if our children at the sound of the school bell learned their lessons in half the time their parents required! We might soon look forward to the day when the ringing of the bells would endow our great grandchildren with all the experiences of the generations that preceded them.[18]

Pavlov, stung by this criticism, asked Studentsov to repeat the experiment with a proper control group (i.e., a group of mice that had not, in a previous generation, been trained to expect food after the sound of a buzzer). The experiment was performed, and to the great disappointment of both Pavlov and Studentsov, no difference in training times between the two groups could be found. Morgan and Bateson were vindicated. Studentsov had mistaken his own improvement as a mouse trainer for inheritance of a learned response by his mice.

After this humiliating event, Pavlov stuck to experiments with dogs, not mice, and stopped praising Studentsov's work. However, according to Pavlov's biographer Daniel Todes, the Russian physiologist never abandoned his belief in the inheritance of acquired characteristics. That faith was a part of his intellectual makeup.

Opposed to "Lamarckists" like Pavlov was a growing school of geneticists in Russia, led by scientists who were thoroughly in step with the development of the field in Great Britain and the United States and who agreed with Morgan and Bateson. This group included N. K. Kol'tsov, A. S. Serebrovskii, I. I. Agol', Theodosius Dobzhansky, S. Chetverikov, Nikolai Vavilov, and Georgi Karpechenko. Kol'tsov, who was a friend of Pavlov even though he disagreed with him, had tried to dissuade Pavlov from talking about the inheritance of acquired characteristics in the United States, but had failed.

However, in revolutionary Russia, geneticists like Kol'tsov suffered from several disadvantages in their debates with the Lamarckists. They were academic geneticists more interested in theory than in practice at a time when Soviet ideologists were demanding that agricultural scientists give practical assistance to Soviet agriculture. Many of the Lamarckists, in agreement with this ideological call, were doing practical work breeding plants and animals. Furthermore, a number of the geneticists were from prerevolutionary bourgeois families who were automatically under suspicion from proletarian radicals.

Thus, long before Lysenko arrived on the scene, the inheritance of acquired characteristics enjoyed a prominence in Soviet Russia that it had lost in preceding decades in the West, especially in Great Britain and the United States. The status of the concept of the inheritance of acquired characteristics in Soviet Russia and its connection with politics is especially evident in the case of Paul Kammerer, an Austrian scientist whose story, detailed in Chapter 3, tells us of the tragic results that can occur when deep political commitments coincide with scientific theory.

3 | PAUL KAMMERER, ENFANT TERRIBLE OF BIOLOGY

He has certainly done a lot of fine things, and he comes uncommonly near showing that an acquired adaptation is transmitted. I don't like it, and shall not give in till no doubt remains. . . . there was something like suspicion of humbug in my mind.

—WILLIAM BATESON ON PAUL KAMMERER

ON THE EVENING of September 22, 1926, the famed Austrian biologist Paul Kammerer checked in to the Rode Hotel in the small resort town of Puchberg near Vienna, the finest hotel in the area (it still exists, though renamed the Schneeberghof). Kammerer had stayed there before and often collected biological specimens, especially salamanders, along the nearby mountain paths. The next morning Kammerer went for a hike along a narrow footpath leading up the mountain behind the hotel. An hour later, he reached a promontory called Theresa's Rock, which offered a dramatic view of the picturesque village below. Over the years and still today, Theresa's Rock has been known as a place where people occasionally commit suicide by throwing themselves off the cliff. Kammerer sat down with his back to the rock, pulled out a revolver, and shot himself. His body sat there, still holding the gun, until a man responsible for maintaining the path found him several hours later.

When the news broke that Kammerer had committed suicide, newspapers in many countries covered the event. After all, this was the biologist whom the *New York World* had called "the greatest of the century" just three years earlier on May 5, 1923. The *New York Times* had hailed him as "the second Darwin."[1] However, six weeks before Kammerer killed himself the leading science publication *Nature* had featured an article by G. K. Noble of the Museum of Natural History in New York accusing Kammerer of fabricating results by injecting india ink into one of his specimens.[2] As a result, Kammerer was totally discredited in what is regarded as one of the greatest scientific scandals of the twentieth century. No biologist of any repute would cite his voluminous works. The only time his name came up was to illustrate how erroneous was the doctrine of the inheritance of acquired characteristics that he so energetically espoused. An exception was Arthur Koestler's 1971 book on Kammerer, in which the author suspected that a right-wing research assistant framed the leftist Kammerer by injecting the ink into his specimen, a midwife toad.[3] However, since Koestler was a novelist, not a biologist, his book did little to redeem Kammerer in the eyes of the scientific establishment. Kammerer, like Lamarck, became closely linked with the concept of the inheritance of acquired characteristics.

Only a change in our understanding of genetics and in particular, the rise of epigenetics, could force another look at the Kammerer affair.[4] Perhaps a change of politics would help as well; Kammerer was a convinced Socialist celebrated in the Soviet Union. In later years his name was often connected to that of Trofim Lysenko, similarly discredited in the West. A thorough reexamination of Kammerer's claims has yet to occur, and he is still largely seen as a dubious scientist. He certainly was an idiosyncratic person, with many human failings. But signs of renewed interest in the case can be seen in the 2009 publication of a news story in *Science* magazine asking if Kammerer's experiments were "fraud or epigenetics?"[5] In that story Azim Surani, a prominent developmental biologist and epigeneticist at the University of Cambridge, is quoted as saying, "It would be

Austrian biologist Paul Kammerer. (George Grantham Bain Collection, Prints and Photographs Division, Library of Congress)

extremely interesting if someone did really try to repeat [Kammerer's] experiment. I wouldn't be surprised if he turned out to be right."[6] And in the same year, Günter P. Wagner, professor of ecology and evolutionary biology at Yale University, called for biologists "to re-examine, with modern molecular techniques, Kammerer's results as a possible case of epigenetic inheritance."[7]

The biological establishment's avoidance of Paul Kammerer's works was illustrated to me personally. In 2013 I checked out his 1924 book, *The Inheritance of Acquired Characteristics,* from the library of the Museum of Comparative Zoology (MCZ) at Harvard University. The MCZ has been the home of some of the most famous specialists in evolution and biological inheritance of the last century, including Ernst Mayr, Steven Jay Gould, Edward O. Wilson, and Richard Lewontin. Ernst Mayr donated his personal library to the MCZ, which is how the Kammerer book ended up there. But Mayr himself never read the book; when I got the book in 2013 many of its pages were uncut and unreadable. A potential reader had held the book in his or her hands and had written on the first page, "Committed suicide 1927 having been shown to have cheated (faked his results) altho the latter is not unquestionably established." (The comment writer had the year wrong—it was 1926.) It is also possible, of course, that Mayr or other scholars at Harvard read Kammerer's book, obtaining copies of it elsewhere. Certainly, they knew about Kammerer and his views (almost all biologists did). The attentive Harvard head of circulation at the MCZ, Ronnie Broadfoot, noticed the pages were uncut and refused to let me proceed with checking the book out, afraid I might damage it by carelessly cutting the pages. The librarian said he would supervise the cutting and told me to return in several hours. I followed his instructions and ended up as the first reader of the Kammerer book from that library in over ninety years.

The book is remarkable and deserves attention at least as a historical document. In our current age, scientific publications are supposed to be devoid of all political and social commentary. Kammerer lived in a different world. The book is divided into two parts, one misleadingly titled "The Biological Part" and the other "The Eugenical Part." Both parts, however, are heavily political, and it is clear that Kammerer believed that science and politics are interwoven. Throughout, he made a case for Lamarckian eugenics (he was strongly opposed to the type of eugenics popular in his day, which rejected Lamarck's views). Even though one might think that the "Biological Part" is mostly scientific, it contains a section titled "Slaves of the Past or Captains of the Future?" that states:

> If acquired characteristics are occasionally inherited, then it becomes evident that we are not slaves of the past—slaves helplessly endeavoring to free ourselves of our shackles—but also captains of our future, who in the course of time will be able to rid ourselves, to a certain extent, of our heavy burdens and to ascend into higher and ever higher strata and development. Education and civilization, hygiene and social endeavors are achievements which are not alone benefiting the single individual, for every action, every word, even every thought may possibly leave an imprint on the generation.[8]

In his section on scientific experiments, only about fifteen pages of his 414-page book are devoted to the midwife toad, the species for which he is most remembered. Instead, it discusses the inheritance of acquired characteristics in a host of organisms, including beetles, butterflies, lizards, chickens, sea squirts, paramecia, caterpillars, polyps, frogs, corn, pine trees, lice, rabbits, and finally, humans. And when Kammerer got to humans, his speculative nature fully emerged. He saw Darwinism (as he defined it, with the inheritance of acquired characteristics an important part) as fully compatible with Socialism and Marxism. According to Kammerer, "real Darwinism" is, "like socialism," a

> doctrine of "upward development" and must concern itself with masses, and not only individuals, or it misses its aim. In the light of such an interpretation, the theory of Natural Selection is not unsocialistic, for its war-cry, "let the best man win," eliminates the prerogatives of birth and money, of *internal* [emphasis added] and external inheritance. Class struggle is a veritable struggle for existence: a race with mental weapons, without violence, a bloodless and a positive selection—the survival of the fittest.[9]

Social Darwinism is usually connected by historians to conservative, even right-wing views, but Kammerer saw it as inherently left wing, even Marxist.[10]

When one sees how Kammerer's deepest political commitments coincided with his support of the inheritance of acquired characteristics, one can easily suspect that he would not be charitable to any scientific evidence undermining that doctrine. Suspicion of his scientific claims seems justified by the very passion with which he presented them and drew on political arguments. But as time passes and the inheritance of acquired characteristics becomes more credible in the twenty-first century, it is not surprising that scientists have begun to suggest that Kammerer might have been right after all.

Kammerer's book is filled with speculations, some extreme. He believed that round-skulled humans have greater brain capacities than long-skulled ones and saw a socioeconomic reason why the lower classes have the disadvantage of long skulls. Impoverished infants are forced to sleep on hard pillows or no pillows at all, while the wealthy have soft pillows that allow their children to develop round skulls and greater brain capacities. All this can be reversed, according to Kammerer, when the poor are able to purchase soft pillows, presumably under Socialism. He even maintained that white "Caucasians" who live long enough as colonists in the Far East develop "slanting eyes" due to the influence of the environment. He grudgingly praised "melting-pot America" even though it is a home of capitalism; more interesting to Kammerer was the fact that the common geographical environment of the American continent, according to him, results in a "new creation" in which European ethnicities are diminished and "certain physical and psychical traits of the American aborigines recur."[11]

We see from all this that Kammerer was a popularizer and a speculator in a way that is alien to scientists seeking reliable evidence and rigorous analysis. It is no wonder that the scientific establishment never accepted his claims about the inheritance of acquired characteristics. And yet he was a skilled experimenter with salamanders, sea squirts, and other organisms; we do not know to this day whether some of these experiments were valid.

Controversies surrounding Kammerer followed him to the little village where he ended his life. In a note found in his pocket, Kammerer asked that his body not be returned to his home in Vienna but disposed of in Puchberg. However, there were difficulties finding a place for him in the local Catholic cemetery. After all, Kammerer was an atheist and a Socialist, and by killing himself he committed a sin according to Catholic doctrine. Puchberg had one cemetery, and the priest was not enthusiastic about burying him there. Finally, he relented but placed Kammerer's body in an area called "suicide corner," without a proper tombstone. Instead, a rough rock with his name attached marked his grave, unfinished and without the usual birth and death inscriptions.

The grave of Paul Kammerer. (Author photo)

Back in Vienna, where Kammerer had many admirers in leftist artistic and literary circles, the news of his undignified internment was disturbing. A group of them commissioned a sculptor to make a large and striking bust of the scientist, which was placed in the garden of the hotel where he spent his last night. It still stands there today. Thus, in Puchberg one can find Kammerer slurred at one end of town and celebrated at the other.

In May 1926, several months before his death, Kammerer visited the Soviet Union.[12] This event sparked a Russian interest in Kammerer that has never totally disappeared. (A Russian tourist agency, MOSINTOUR, is currently offering a "Paul Kammerer Tour" that features visits to places in Europe where genetics was being developed between the wars.)[13] While Kammerer was in Moscow, he met with the Soviet minister of education, Anatoly Lunacharsky. Lunacharsky was a sophisticated, well-educated man deeply interested in culture and science. He would later be replaced by much rougher characters. In his memoirs Lunacharsky described his encounter with Kammerer:

> On one wonderful day a tall man with an open and intelligent face came to my office. I already knew quite a bit about Professor Kammerer; I knew that he was one of the great partisans of the theory of the inheritance of acquired characteristics in Europe, and I also knew that he was a great friend of Soviet Russia and that he was proposing to transfer the center of his laboratory research to Moscow. We had a long and substantial conversation.[14]

A topic of conversation was why the theory of acquired characteristics encountered such resistance among Western geneticists. Kammerer, no doubt trying to win favor with the man he hoped

The author with a bust honoring Paul Kammerer. (Author photo)

would finance his research in Russia, described this resistance in Marxist terms:

> The hostility toward the theory is, in my opinion, exactly what you say, a class hostility. If all the development of the plant and animal world is unrelated to the external environment, and occurs only as a result of some mysterious selection, that is by means of changes in the germ plasm that are secret to us, then there remains a big place for the tired science of "Providence."[15]

Kammerer explained that a Socialist society like Soviet Russia had an opportunity to "change all humankind in the future" by improving living conditions and constructing a "correct" educational system that would have heritable effects.

Lunacharsky was fascinated by this vision. A form of Lamarckian eugenics of the type Kammerer described was still possible in the Soviet Union of the 1920s. Only later, after the rise of both Stalinism in Russia and Fascism in Germany, would Soviet ideologists condemn eugenics of any kind. Lysenko would join in this castigation of eugenics, even one based on his theories. But in the 1920s, the idea of genetics applied to human society was still attractive to many Soviet intellectuals.

Kammerer described in graphic terms to Lunacharsky the vision of society according to the neo-Darwinists in Western Europe:

> From the point of view of Neo-Darwinism, society is like a container of milk in which the cream has risen to the top—the aristocracy, the great bourgeois companies, the intellectual mandarins with "inherited intellectual abilities."

And suddenly Revolution is like an enormous spoon which mixes all this up.[16]

In his recollection of this meeting, Lunacharsky admitted that he knew little about biology and assured himself that Communists "will not falsify science because some conclusions might be unpleasant for the proletarian class."[17] Nonetheless, he said he had "unlimited sympathy" for Kammerer's point of view.

The meeting stuck in Lunacharsky's mind, and several months later when he heard about Kammerer's suicide he returned to it. He also noticed what he saw as an effort in the West to discredit not only Kammerer's science but also his person by spreading "rumors about the dark sides of Kammerer in social, family and other aspects."[18] And it is certainly true that Kammerer was a womanizer and a libertine. Even his daughter Lacerta (named after one of his favorite lizards) acknowledged this but generally blamed the women he got involved with:

> He was so used to the magnetic effect he had on women that he did not try to make conquests—with very few exceptions; he conquered too many for his own comfort as it was. . . . And did he attract them! I remember streams of them trying to get "in his presence" with all sorts of excuses.[19]

An example of Kammerer's social unsteadiness is his relationship with Alma Mahler, the widow of the composer Gustav Mahler. Alma Mahler was a femme fatale in Vienna who successively became the wife of Gustav Mahler, the architect Walter Gropius, and the novelist Franz Werfel and had affairs with a number of other prominent men, including the artist Oskar Kokoschka. In 1911 Alma became a laboratory assistant to Paul Kammerer in his Vienna laboratory. A tumultuous and passionate love affair followed. Kammerer proposed marriage and swore that if Alma did not accept he would shoot himself on Gustav Mahler's grave. Alma called his bluff; Kammerer backed down and careened on to other adventures.

Lunacharsky in Moscow heard about Kammerer's antics but was not offended. Instead, he saw a great propaganda opportunity. He loved the artistic process of creating movies and had already authored several film scripts, including the popular melodrama *The Bear's Wedding*.[20] He conceived of a motion picture in which he and his wife starred as saviors of the maligned socialist biologist. He worked with *The Bear's Wedding* screenwriter Georgii Grebner. He wanted to use the film to praise science but at the same time castigate malevolent forces such as the Catholic Church, the bourgeoisie, and rising Fascism. He wrote in his memoirs that he "imagined a film of the following sort":

> In the center is a professor, an attractive talented scientist, a man who was deeply progressive and connected with the working class. . . . The Church promoted a patriotic party of a fascist character which had strong branches in the student body of the university. The Church was represented in the university by a priest who was at the same time a professor—a theologian occupying a prominent place in biological science. He was the chief enemy of the professor-revolutionary and was the initiator of a complicated conspiracy dedicated to destroying him as soon as possible. . . . Fascism was personified in a local prince who I decided to copy from the well-known Prince Lichtenstein, an aristocrat, a convinced Catholic, a gambler, a bonvivant, and a counterfeiter known throughout the world.[21]

Lunacharsky had previously worked with a left-wing German film producer, Prometheusfilm, and he again agreed to have them produce the film.[22] The Germans chose the city of Erfurt, with its ancient cathedral and medieval center, as the main locus for the film, although some scenes were shot in other German cities. The director of the film was the Russian Grigorii Roshal', a well-known film-maker who in his long career directed over two dozen Soviet films. The main star of the film, Bernhard Goetzke, playing the role of the professor (Kammerer), was one of the best-known silent film stars in Germany, although about to go into decline with the advent of sound films. Lunacharsky's *Salamandra,* a silent film, was one of his last works.

In the film the conspiracy against the professor works. His enemies surreptitiously inject ink into one of the professor's salamanders, allegedly showing spots that are inherited acquired characteristics. They publicly place the salamander in a solution, washing out the ink and exposing the professor as a fraud. The film deviates from the truth in many ways, perhaps most dramatically in its denouement. Instead of committing suicide, the ostracized professor is saved by a former student who contacts Moscow for help. The professor then receives a phone call from none other than Soviet minister of education Lunacharsky, inviting him to Moscow to continue his scientific experiments justifying the inheritance of acquired characteristics. The final scene shows the professor and the woman who saved him joyfully journeying on the train to the Soviet Union, which is described as a country "where geniuses are valued."

The film became a hit in Russia in 1929 but was also shown in Germany. The film outraged Fritz Lenz, a German biologist and exponent of genetic racism who would become a member of the Nazi Party. He devoted eight pages in his journal to a sharp review of it. "Everyone involved with Kammerer was Jewish, Bolshevik, and ma-

Soviet commissar Anatoly Lunacharsky and his wife, Natalya Rozenel, 1927. (Georg Pahl/Bundesarchiv)

liciously motivated by problematic politics," he maintained, adding that Kammerer was "half-Jewish." He said that Lamarckism particularly represented the Jewish fantasy that "by living in the German environment and adapting to German culture, Jews could become true Germans."[23]

A genetics conference was held in Leningrad in 1929 while the film was in local theaters. Most of the attendees were Russian, but a few foreigners were present, including prominent geneticist Richard Goldschmidt. Goldschmidt was at the time one of those Jews living in Germany to whom Lenz referred. Not surprisingly, Goldschmidt and Lenz were not friends.[24] Goldschmidt, at this moment a researcher at the Kaiser Wilhelm Institute of Biology in Berlin, would flee the anti-Semitism in Germany in 1935 and emigrate to the United States, where he would become a distinguished scientist at the University of California, Berkeley.

As different as Goldschmidt's politics was from that of Lenz, he also criticized Kammerer. He reminisced about his experience at the Russian genetics conference:

> One day, walking down the street with my friend Philip-chenko [*sic*], I saw in front of a movie house a large poster of *Salamandra* decorated with pictures of this harmless animal. My surprised question was answered by my friend with an invitation to see the film. This we did, and my friend interpreted the text. . . . [The film] turned out to be nothing but a propaganda film for the doctrine of the inheritance of acquired characters. It uses the tragic figure of Kammerer, his salamanders, and mixed up with them, for the story, his midwife toads.[25]

Thus, although Kammerer was in complete disgrace in Western biology, sympathy for him continued for decades in Russia. A trace of that sympathy can still be found. The current Russian *Meditsinskaia entsiklopediia* (Medical Encyclopedia) contains an article on Kammerer that asserts his innocence.[26] Although Kammerer enjoyed nowhere near Lysenko's influence on Russian science, the country's sympathy for him reflects some of the same factors that have prompted a revival of Lysenkoism.

4 | THE GREAT DEBATE ABOUT HUMAN HEREDITY IN 1920S RUSSIA

> The understanding of the heredity of acquired characteristics brings us a message of salvation. . . . We may assume the deeds and thoughts of man may be passed on to future generations, resulting in a "man sublime."
> —PAUL KAMMERER

WHEN PAUL KAMMERER ARRIVED in Russia in 1926, he found his ideas about the inheritance of acquired characteristics in the middle of a great debate about human heredity and its meaning. Although pleased that his views were more appreciated in Russia than in Western Europe, Kammerer was surprised and even dismayed by the complexity of the discussions. So many different viewpoints were being propagated that it was difficult to follow them all. Often, Russians took Kammerer's belief that politics and science are interwoven to a degree he had never dreamed. Understanding these debates is necessary to fathom the soil from which Lysenko emerged.

One rather small group of non-Marxist Mendelian, or classical, geneticists such as N. K. Kol'tsov denied the inheritance of acquired characteristics and wanted to implement eugenics in a way that was rather popular in Western Europe and America. Kol'tsov emphasized the independence and significance of human genes in favoring specific traits. In Kol'tsov's view, the social and political environments were irrelevant to human heredity; genes were what counted. Kol'tsov's supporters debated about how to apply this Mendelian eugenics: Should it be used in positive ways (the encouragement of "good" sexual matings) or negative ways (the prevention of "bad" sexual matings)?

Then there were the Marxist classical geneticists, both foreign and native, such as the American H. J. Muller (a frequent visitor) and the Russians A. S. Serebrovskii and S. N. Davidenkov. These Marxist geneticists were also eugenicists but believed that eugenics could be properly applied only in a Socialist society in which economic inequality and privilege had been eliminated. They agreed with Kol'tsov on the new genetic principles being developed largely in Britain and America. They also, like him, denied the inheritance of acquired characteristics. But they disagreed with Kol'tsov on political grounds and saw him as a remnant of bourgeois culture who would favor his own kind. Who would decide what was "good" and "bad" in eugenic choices? The Marxist geneticists did not wish to give that power to the likes of Kol'tsov, who came from a pre-Revolutionary merchant family and showed little political loyalty to the Bolsheviks now in power.

The Marxist biosocial eugenicists, comprising a third group, provided yet another viewpoint in Soviet Russia. They believed that a new type of Lamarckian "biosocial eugenics" should be applied and placed the inheritance of acquired characteristics at the center of their attention. They included Kammerer himself and Russians such as S. G. Levit, E. S. Smirnov, Iu. M. Vermel', B. S. Kuzin, and M. Volotskoi. They were convinced that creating a society in which education, health care, and economic sustenance were freely available to all would benefit not only the citizens of the new Soviet Union but their progeny as well. These descendants would inherit the positive acquired characteristics of their parents living in a Socialist society. The Soviet Union could literally produce "A New Soviet Man" who differed from his capitalist counterparts not only politically but genetically.

Yet a fourth group, containing Marxists such as Vasilii Slepkov, many of whom were not geneticists, believed that the whole idea of making any type of genetics a key to Soviet society was ridiculous and anti-Marxist. They held that Soviet society should be governed by the political and economic principles of Marxism, not the principles of a natural science like genetics. Let us now look at each of these groups more carefully.

THE MENDELIAN GENETICISTS

Nikolai Kol'tsov (1872–1940), who became the leader of the classical Mendelian geneticists in Russia, was born in Moscow to a well-off family. He studied comparative anatomy and comparative embryology at Moscow University. Shortly after receiving his master's degree, Kol'tsov worked on a two-year foreign study grant in laboratories in Germany, Italy, and France. Research there on cellular biology resulted in further graduate study at Moscow University, where he taught for a number of years.

In his politics Kol'tsov was a liberal democrat critical of the tsarist regime, participating in protests before and after the 1905 Revolution. After the overthrow of tsarism in 1917, Kol'tsov organized the Institute of Experimental Biology (IEB), where he pursued his interests in genetics and eugenics. Shortly thereafter, he initiated the Russian Eugenics Society, of which he was president. The society published the *Russian Eugenics Journal* from 1922 to 1930.

Kol'tsov's enthusiasm for eugenics and his opposition to the doctrine of the inheritance of acquired characteristics were striking. He maintained that "eugenics is the religion of the future, and awaits its

Russian genetics pioneer Nikolai Kol'tsov. (RIA Novosti/Science Source)

prophets."[1] At the same time, he thought of himself as one of those prophets. He considered eugenics to be a subdiscipline of "zootechnics," the science of the improvement of animal breeds.[2] To illustrate the similarity between the breeding of animals and the breeding of humans, he once fantasized about the enslavement of humans by extraterrestrial invaders who conquered Earth. By following a prescribed breeding program, these extraterrestrials, Kol'tsov wrote, could "in as little time as a century create endless individual races of domesticated people as sharply distinct from one another as a pug or a lapdog is from a Great Dane or St. Bernard."[3] Of course, such a vision offended Kol'tsov, who insisted that one of humanity's precious freedoms was "the right to freely choose a spouse," but he hoped that with sufficient knowledge of eugenic principles, such choosing could be voluntarily done in a way that would improve the human race.

The historian who today leafs through the pages of Kol'tsov's *Russian Eugenics Journal* is struck by its authors' naiveté and blindness to political complications. In the years immediately after the revolution, they were concerned with the genealogy of outstanding and aristocratic Russian families; investigations, complete with family tables, were made of these families, as well as all the members of the Academy of Sciences in the previous century (Kol'tsov became a corresponding member of that academy during the tsarist period). Several authors expressed dismay about the dysgenic effects of the Russian Revolution. Kol'tsov lamented that in the French Revolution "the flower of the French nation went to the guillotine."[4] The emigration of the nobility and of other upper-class families after the Russian Revolution of 1917 was seen as a serious loss to the genetic reserves of Russia, requiring eugenic correction.

More predictable, perhaps, was concern about the enormous Russian losses during World War I, far heavier in absolute (though not in relative) terms than those of any other nation. One Soviet eugenicist, V. V. Bunak, stated that the total genetic impact of eight years of war, revolution, civil war, and famine "might exceed that of the West" in the same period. Thus, the deep eugenic gloom of postwar Germany was reflected among Russian eugenicists, although there was some hope for backward Russia in that "the more cultured a country, the more the biological danger of war."[5] Kol'tsov, holding out no hope that environmental influences could alleviate this damage, said that Kammerer's belief in the inheritance of acquired characteristics was denied by "the modern theory of heredity."[6]

The favored word in the *Russian Eugenics Journal* for the field was *evgenika* (eugenics), although the term *rasovaia gigiena* (race hygiene) was also used, and racial interests were widespread. The Russian Eugenics Society established a special commission for the study of the "Jewish race," following a major interest in the German movement. In one study the commission concluded that Jews were in no way inferior to other ethnic groups.[7] Despite the conclusion the Russian eugenics movement explicitly thought in terms of racial differences, psychological as well as physiological. Kol'tsov was impressed by the German guidelines of race hygiene in terms of a static or typological definition of race (such as the infamous textbook on human heredity by Baur, Fischer, and Lenz).[8] The first issue of the Russian journal contained reviews of fourteen German books on human heredity and eugenics—and no others.

The leaders of the Russian eugenics movement were anxious for contacts with other national eugenics societies and worked for full international recognition.[9] The fact that Soviet Russia was attacking the practices and worldviews of leading nations in most areas of international political and cultural relations seems not to have deterred Kol'tsov and the eugenicists around him. They established contacts with the Eugenic Education Society in England, the Eugenic Record Office in the United States, and the German Society for Race and Social Biology. In November 1921 the Russian Eugenics Society was recognized as a full member of the International Eugenics Union.

But in the international arena, Bolshevik Russia (as well as Germany) was ostracized after World War I by the former Allies at many international scholarly congresses. Thus, neither Soviet Russia nor Germany was invited to the second International Congress of Eugenics in New York in 1921, a slight that annoyed Kol'tsov. This ostracism only deepened scholarly connections between the two pariah nations.

Kol'tsov managed to obtain an invitation to the Third International Congress of Eugenics in Milan in 1924, where he lamented that the Catholic Church was so influential that the participants were one-sidedly cautious in their discussions of practical eugenic measures. Kol'tsov noted that if the attitudes of Catholic Italians were one extreme, those of the Americans, with their system of sterilization laws, were another. No Americans or Germans were present at the congress, and according to Kol'tsov, not one participant spoke in favor of the American system of sterilization. However, in the same year Kol'tsov's journal gave publicity to a German physician who appealed to his "colleague-physicians . . . to search for defective persons . . . and to sterilize as many of them as possible."[10]

Although Russian eugenicists such as Kol'tsov shared many assumptions of the international eugenics movement in general and the German movement in particular, it would be an exaggeration to see them as proto-National Socialists. Substantive links between the Russian eugenicists and incipient Fascism were lacking. Similar eugenics organizations existed in most Western nations at that time and in fact were far stronger in England and in the United States than in Russia. Nonetheless, it is remarkable that Kol'tsov and his fellows went as far as they did in embracing radical eugenic measures based on prejudicial definitions of human worth. It was inevitable that they would get in trouble with a new revolutionary Soviet state based, at least in principle, on preference for the lower-class proletariat.

MARXIST MENDELIAN GENETICISTS AND EUGENICISTS

Marxist geneticists and eugenicists in Russia in the 1920s differed sharply with many of Kol'tsov's assumptions and conclusions, especially regarding the dysgenic effects of the Russian Revolution. They wanted to create a totally new kind of eugenics, one that would help Soviet Russia. Among them were H. J. Muller, A. S. Serebrovskii, and S. N. Davidenkov.

American Nobelist H. J. Muller, in suit and tie. (American Philosophical
Society/SPL/Science Source)

Muller, who would win the Nobel Prize in 1927, came to the
Soviet Union in 1922, bringing with him samples of the fruit fly *Dro-
sophila melanogaster* from the famous laboratory of Thomas Hunt
Morgan at Columbia University. That laboratory did pioneering
work on the study of chromosomes. Muller, a Marxist, a sympathizer
with the Soviet Union, and an avid eugenicist, wanted to help Russian geneticists apply this important research toward creating a new
type of Soviet citizen. Muller was passionate about the possibilities
of this type of eugenics in Soviet Russia, which were not present, he
believed, in the capitalist United States.

A. S. Serebrovskii, one of Muller's friends, also shared Muller's
eugenic views. In an article titled "Anthropogenetics and Eugenics

in a Socialist Society," Serebrovskii criticized Kol'tsov for maintaining that eugenics should be a "religion," seeing this as a typical "bourgeois" interpretation.[11] Serebrovskii lamented the fact that eugenics is the "daughter of bourgeois parents" when "it is quite clear that only a socialist society can provide a good home and upbringing for this discipline."[12]

How, Serebrovskii asked, should eugenics be applied in the Socialist Soviet Union? Only, he said, by realizing that the destruction of private property in the economy should be extended to the destruction of the family itself, another form of private property. He observed, "A conceited bourgeois property owner recognizes only his own children. His wife must give birth only to his children."[13] On the other hand, under Socialism, postulated Serebrovskii, "Love will be separated from childbirth," and "sperm must be obtained from a specific approved source."[14] He continued:

Because man possesses an enormous sperm-producing capacity, using the excellent modern technique of artificial insemination . . . it is possible to obtain from one outstanding and valuable breeder up to 1000 or even 10,000 children. Under such conditions, human selection will make giant leaps forward. Then individual women and entire communes will take pride not in "their own" children, but in the successes and achievements in this most remarkable field—the field of creation of new forms of humans.[15]

S. N. Davidenkov shared Serebrovskii's political allegiances and his devotion to eugenics in a Socialist society, but he put less emphasis on artificial insemination as a means of creating a New Soviet Man. He thought the same goal could be achieved by making a "mandatory eugenic survey of the urban population of the Soviet

Soviet geneticist Aleksandr Serebrovskii. (RIA Novosti/Science Source)

Union" and then favoring the "right" marriages with "state compensation of child-related expenses through the proportional increases of salaries." The most intellectually gifted parents would be encouraged to have more children by increasing their salaries by 50 percent with every childbirth.[16]

Later, H. J. Muller would actually propose eugenic measures of the sort favored by Serebrovskii and Davidenkov in a letter to the leader of the Soviet Union, Joseph Stalin. He opened his letter with the sentence, "As a scientist with confidence in the ultimate Bolshevik triumph throughout all possible spheres of human endeavor, I come to you with a matter of vital importance arising out of my own science—biology, and, in particular, genetics." Muller then went on to propose that the sperm from "the most transcendently superior individuals, of the 1 in 50,000" should be artificially inseminated into Russian women for the birth of a higher type of human being. He gave as examples of these "transcendently superior individuals" Lenin and Darwin.[17]

Stalin had by this time discarded visions of a radical transformation of Soviet society and had moved toward greater attention to religion, established social norms, and patriotism in the face of the looming threat of Nazi Germany. He totally rejected Muller's appeal.

MARXIST, OR BIOSOCIAL, EUGENICS

Paul Kammerer was an outspoken supporter of a completely different kind of eugenics from that supported by Mendelian geneticists. For Kammerer, the social environment was of primary importance. He believed that the creation of beneficial social and economic conditions would change the heredity of Soviet citizens forever through the inheritance of favorable acquired characteristics. Only in that way, he maintained, could Soviet Socialism produce a superior type of human being—a citizen who differed hereditarily from his or her capitalist counterparts.

Kammerer had considerable influence in Soviet biological circles. His research and writings were well known. In Leningrad he attended a tumultuous meeting of the Society of Materialist Biologists where one of the speakers exulted at his appearance and said it was "an important event for us Lamarckians." Lamarckian views were widespread among Russian biologists, especially young Marxists but also many non-Marxists.

Kammerer preached that "the understanding of the heredity of acquired characteristics brings us a message of salvation."[18] Through this doctrine, he continued, "We may assume that the deeds and thoughts of man may be passed on" to future generations, resulting in a "man sublime."[19] He called for a "productive eugenics," which he contrasted sharply with the "prejudicial" Mendelian eugenics of Kol'tsov. Kammerer had many supporters among the participants in the great debate in Russia over problems of heredity, including S. G. Levit, E. S. Smirnov, Iu. M. Vermel', B. S. Kuzin, and M. Volotskoi. One of the most colorful was Levit.

Solomon G. Levit (1894–1938) was born to a poor Jewish family in Lithuania.[20] Aided by his unquestioned academic talent, he managed to be admitted to Moscow University, where he received a medical education and became a researcher and teacher, particularly on the topic of diseases of the blood. In 1920 he joined the Communist Party and eventually became active in the Marxist "Circle of Materialist Physicians."

In a 1924 paper titled "Marxism and Theories of Biological Evolution," Levit strongly supported Lamarckism, declaring that the "general force" directing the course of evolution is "the influence of the environment."[21] He believed that he could combine Lamarckism and Darwinism without contradiction. He stretched that belief by asserting that "Lamarckian factors (the influence of external factors, the inheritance of acquired characteristics) are virtually the only ones capable of explaining the causes of variation."[22] In a 1926 paper, he maintained that "the Soviet proletariat—with the majority of Soviet physicians not far behind—accepted the inheritance of acquired characteristics long ago." He described advocates of Mendelian genetics as "pessimistic and impotent."[23]

Levit's views were echoed by M. Volotskoi, who openly called for a "biosocial eugenics" that would emphasize education and social reforms in changing the heredity of Soviet citizens through the inheritance of acquired characteristics. He wrote, "The eugenics developed in bourgeois countries (with which proletarian eugenics can have nothing in common) has always striven to obscure every new discovery which favors this type of inheritance."[24]

The Russian Mendelian geneticists, such as Kol'tsov and his colleagues, were very upset by this sort of argument. In contrast to the Marxist supporters of the inheritance of acquired characteristics, they had long ago, along with their mentors Bateson and Morgan, dismissed the doctrine as nonsense. Iurii Filipchenko, a member of this group, rose to the challenge of opposing Kammerer and other biologists on this issue. He is still remembered today as one of the authors of the "modern synthesis" bringing Mendelian genetics and Darwinian evolution together.

Filipchenko, a passionate opponent of the theory of the inheritance of acquired characteristics, expressed this view in an article published together with a translated article by Morgan on the same subject (a method of presentation sure to infuriate Marxist Lamarckians).[25] Filipichenko tried to turn the tables on these Marxists by making the following argument: he said that the Marxists were assuming that only "good" environments have hereditable effects,

while a consistent interpretation of the inheritance of acquired characteristics would show that "bad" environments also have effects.
Therefore, all socially or physically disadvantaged groups, races, and
classes of people, such as the proletariat and the peasantry and the
nonwhite races, would have inherited the debilitating effects of living
for centuries under deprived conditions. Far from promising rapid
social reform, the inheritance of acquired characteristics would mean
that the upper classes would be not only socially and economically
advantaged but also genetically privileged as a result of centuries of
living in a beneficial environment. Thus, the proletariat in Soviet
Russia would never be capable of running the state; it was genetically
lamed by the inherited effects of its poverty. Filipchenko's recommended exit from this dilemma was simple: give up the whole idea of
the inheritance of acquired characteristics.

The critics of Mendelian geneticists hesitated before the logic of
Filipchenko's argument. Several radical journals ran articles maintaining that only the inheritance of acquired characteristics, not
Mendelian genetics, was counterrevolutionary. One author stated that
the international bourgeoisie constantly renewed efforts to establish
the inheritance of acquired characteristics in order to show its genetic superiority, but the proletariat was learning that science spoke
against them.[26] Bourgeois professors from the West, like Paul Kammerer, should not be trusted on this issue. Another author, writing
in the *Red Journal for All People,* said that every social reformer should
read Filipchenko's argument in order to be armed for the political
struggle.[27]

M. V. Volotskoi, an ardent defender of Marxist biosocial eugenics,
was not willing to accept Filipchenko's argument.[28] He maintained
that Filipchenko's mistake lay in believing that the genetic effects of
the inheritance of acquired characteristics would be unilaterally
harmful to those from a background of exploitation and unilaterally

Soviet entomologist and geneticist Yurii Filipchenko, 1929. (American
Philosophical Society/SPL/Science Source)

useful to those from a privileged background. Marxists understood, said Volotskoi, that the division of labor that led to societies based on slave owners and slaves, lords and peasants, or capitalists and workers has harmful and beneficial effects for both groups.

On each side of exploitative relations, said Volotskoi, there is a balance sheet of pluses and minuses: the proletariat suffers from poverty, lack of culture, and unsanitary conditions, but it benefits from the physical labor and hard work necessary for survival; the capitalists benefit economically from their privileged position, but they do not have to labor and therefore become slothful and corrupt. As a result, Volotskoi continued, an assumed inheritance of acquired characteristics does not only benefit the upper classes; its effect would be mixed. And any harmful effects that did occur could be erased in just a generation or two. That erasure would come, Volotskoi was convinced, by living in a just and beneficial Socialist society.

The conflict between Volotskoi and Filipchenko and by implication, between all proponents of biosocial eugenics, on the one hand, and Mendelian eugenics, on the other, had reached a stage in which anybody could manufacture arguments in favor of any point of view, given the will. This stalemate prepared the way for the next group of partisans, who maintained that human genetics was irrelevant to the task of building a Socialist society.

SOCIAL THEORY OVER BIOLOGY

Vasilii Slepkov was a Marxist who thought all this talk of eugenics and biology as keys to the Soviet future had gone too far. He published an article in the major Bolshevik theoretical journal in which he said that the eugenicists—of whatever type, Lamarckian or Mendelian—were emphasizing biological determinants of human behavior while totally neglecting socioeconomic determinants.[29] Since most eugenicists were biologists with little knowledge of the social sciences and even less of Marxism, they had "absolutized" the influence of heredity to the detriment of Marxist historical materialism and economic determinism. They had converted genetics into a "universalist" interpretation in which all of human history was merely a replacement of one genotype by another. Slepkov pointed out that this interpretation totally ignored the principles of Marxism, which demonstrated that social conditions determine consciousness. Slepkov quoted Karl Marx's concept that "people are a product of conditions and education and, consequently, changing people are a product of changing circumstances and different education."[30]

A thief is not a biological type created by heredity, said Slepkov, but a "social man" created by his environment, poverty, and unemployment. Slepkov did not deny, however, that biological organisms differed genetically, and if he had to choose between Lamarckian and Mendelian genetics, he definitely preferred the Lamarckian approach. But although the inheritance of acquired characteristics may be the way to explain plants and animals, he indicated, it was not the way to explain humans. For this latter task, political and economic Marxism was the correct approach.

Slepkov's article opened up the whole question of the relationship of natural science to Marxism as a political theory. Both natural scientists and Marxists often called themselves "materialists," believing that all that exists is matter, but what did this mean exactly for biology, and even more importantly, for society? Was "matter" the same on all levels—the inanimate matter of physics and chemistry, the living matter of biology, and the social matter of human civilization—or were there "different scientific laws on different levels of being?"

Friedrich Engels, the fervent friend and promoter of Karl Marx, had said that Marx discovered the "law of evolution of human history" just as Darwin had discovered "the law of evolution of organic nature."[31] Engels seemed to be saying that although Darwin and Marx used similar methodologies to study different realms, the realms are indeed not the same, and therefore the Marxist "laws" of society cannot be reduced to the biological "laws" of living matter. Did that mean that Kammerer's beloved doctrine of the inheritance of acquired characteristics, based on salamanders and toads, should not be directly applied to humans? Maybe human actions were determined by social science, especially Marxism, and not biology? When Kammerer arrived in Russia in 1926, these questions were at the heart of the controversies surrounding him.

AN OFFICIAL CONSENSUS EMERGES

The variety of scholastic arguments about the relationship of biology to society bewildered Kammerer, who had a rather simple vision. He was convinced that good Marxist Socialists should favor the inheritance of acquired characteristics. He hoped, even assumed, that the Soviet Union would embrace this doctrine and use it to build a new society in which its citizens benefitted from favorable economic and social conditions—not only individually in one generation but also biologically in subsequent ones.

Kammerer was caught off guard when he learned that some militant young Marxists looked upon him as a "bourgeois Western professor" despite his enthusiasm for Soviet Socialism. At a time when wearing a necktie was a symbol of political sympathies, the nattily attired (he liked bow ties) Kammerer was not immediately welcomed by all radicals.[32] It may have been at this time that Kammerer decided that moving his laboratory to Russia, as Lunacharsky had invited him to do, was not entirely wise.

Out of this confusing maelstrom of ideas and politics in early Soviet Russia, however, an official consensus emerged—one that was enforced by the Communist Party. It held that Mendelian genetics was a suspicious field and definitely should not be applied to human beings in the form of eugenics. Lamarckian biology was favored but only on the levels of plants and animals, not humans. Humans were explicable in Marxist terms, not biological ones.

These principles had gained great currency in Soviet Russia long before Lysenko became influential in the mid-1930s. But when he arrived on the scene, he subscribed to them, concentrated on plants and animals, and eschewed all talk of the relevance of biology to human beings. And he worked to get rid of his opponents.

If Kammerer had stayed in Soviet Russia, he would have eventually found his views rejected and his own personal safety in jeopardy, especially after the rise of Fascism in Germany with its attention to genetic race hygiene. Kammerer's protestations (in German) that he favored a different kind of eugenics, not the Nazi type, would not have carried much weight. Kammerer would have been seen as a biologist more interested in biological "laws" than Marxist "social laws." Many biologists would be arrested in the coming years, and some would be Lamarckian eugenicists like Kammerer.

The two foreigners who participated in the great debate about human heredity in Russia in the 1920s—the Austrian Kammerer and the American Muller—were in the end rebuffed by their hosts. Both sympathized with the Russian Revolution, and both hoped to help it. Although they preached dramatically different forms of eugenics, the Russian Communists eventually turned against them. Kammerer returned home to an unreceptive biological community in Europe, where he decided to end his life. Muller returned home to an unreceptive political community in the United States and had to answer for his Communist sympathies.[33]

5 | LYSENKO UP CLOSE

You think I am a part of the Soviet oppressive system. But I have
always been an outsider.... I had to fight to be recognized.

—TROFIM LYSENKO

In 1971 I was in Moscow, doing research on Lysenko and feeling
frustrated. The man was still alive, but all my attempts to interview
him had been unsuccessful. I had first tried to contact him ten years
earlier, when I was a student at Moscow University. Then, he had been
fully in power, dominating Soviet biology.

At that time, from the top of the skyscraper at the center of the
university, I could see Lysenko's large and well-equipped Lenin Hills
Farm, where he conducted experiments with milk cows, trying to
increase their production on the basis of the inheritance of acquired
characteristics. I went to the farm's administrative office and left
copies of articles I was writing about him, with a note attached pro-
viding my phone number. I wrote that these manuscripts would soon
be published in the West but added that I still had time to make
changes if he would meet with me. I was counting on his ego, hoping
he would want to try to influence my work. He never answered. Ten
years later, in 1971, after he had been discredited in the Soviet Union, I
repeated my efforts, this time leaving new manuscripts and my contact information at his Academy of Sciences office (he was deposed
as the tsar of Soviet biology in 1965 but still retained his prestigious
position at the academy). Again, the result was the same. He did not
wish to see me.

So I gave up. With no chance to interview him, I turned entirely to libraries and archives, where I found voluminous information about him. I read everything I could about Lysenko and learned the history of his professional life and his academic writings intimately. I spent months working at one of the best libraries for this purpose, the Lenin Library in downtown Moscow. The lunch facilities in the basement of the library were so abysmal that I sought better food and a more attractive setting elsewhere. One of the best spots, the House of Scientists a few blocks away, was a sort of faculty club belonging to the Russian Academy of Sciences. Since I was in the Soviet Union on an official exchange program between this academy and the National Academy of Sciences of the United States, I had a pass that would permit me to use all the Russian academy's facilities.

The House of Scientists was a richly ornamented prerevolutionary building. Originally built by a nobleman in the eighteenth century, it had been badly burned in the fire of 1812 during Napoleon's occupation of the city. Rebuilt in the nineteenth century, it became one of the most luxurious and famous meeting spots of the nobility and the richest merchants in Moscow. At one time or another it was occupied by relatives of Peter the Great's mother (the Naryshkins), the relatives of Russia's greatest poet Alexander Pushkin, or the famous composer Nikolai Rimskii-Korsakov. It was frequented by internationally known writers such as Ivan Turgenev and Nikolai Gogol. In the latter part of the nineteenth century, it fell into the hands of an industrialist and merchant family, the Konshins. After the Russian Revolution of 1917, the building was confiscated by the victorious Communists and converted into a magnificent social spa for the scientists of the Russian Academy of Sciences. They were to be the new nobility.

On this early spring day in 1971, I entered the House of Scientist's palatial dining room after spending the morning in the Lenin Library. Sitting alone at a table at the back of the room was a gaunt and homely man. I immediately recognized Trofim Lysenko. In the Soviet Union, it was not unusual for strangers to sit at the same lunch table, so I sat down beside Lysenko, ordered a bowl of borscht, and began to eat.

After a while I turned to Lysenko and said, "I know that you are Trofim Denisovich Lysenko. I am Loren Graham, an American historian of science, and I have written quite a bit about you. Several times I have sent you my work." Lysenko replied, "I recognize your name. I have read what you wrote about me. You know a lot about Russian science, but you have made several serious mistakes in describing me and my work."

I immediately inquired what my mistakes were. Lysenko answered:

> The most important mistake was that you accuse me of being responsible for the deaths of many Russian biologists, such as the well-known geneticist Nikolai Vavilov. I disagreed with Vavilov on biological issues but I had nothing to do with his death in prison. You know, I have never even been a member of the Communist Party, and I am not responsible for what either the Party or the secret police did in biology.

I was silently grateful that I had spent the previous months in libraries and archives acquiring a great deal of information about Lysenko and his victims. I knew he was correct in saying that he was not a member of the Communist Party, a fact I had cited in my previous publications. But he erred grossly in saying that he bore no re-

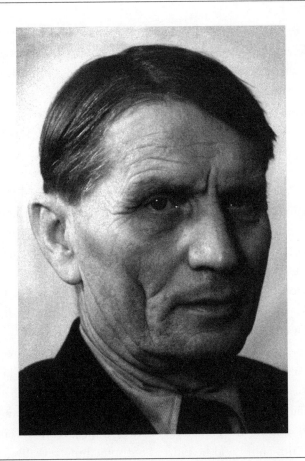

Lysenko and his work fell into disgrace in 1965, though he kept his
position in the Soviet Academy of Sciences until his death in 1976.
(RIA Novosti/Science Source)

sponsibility for the deaths and imprisonment of prominent Soviet
geneticists. His method was lethal and passive-aggressive. He portrayed himself as a simple agronomist—even a peasant—who had a
successful agricultural method the establishment geneticists would
not accept. Yet he had described the leading academic geneticists as

traitors to the Soviet cause, counterrevolutionaries and foreign agents purposely harming Soviet agriculture. And in so doing, Lysenko attracted the secret police's attention to them. He did this many times. But when the police arrested his critics as "traitors," he claimed he had nothing to do with the arrests.

Lysenko's method, called "denunciation" *(donos),* was actually well known in Soviet times. Many people knew it was possible to get the secret police to rid one of enemies or rivals by denouncing them as "anti-Soviet" or "treasonous." People often used denunciations against a professional competitor, a rival in a love triangle, or a political opponent. Denunciations could be oral or written; in Lysenko's case they were oral. Such acts usually had double effects: they successfully eliminated the rival but implicated the denunciator in the Soviet system. Lysenko, however, refused to recognize that implication. Or, at least, he refused to publicly admit his guilt, which was clear enough to many others.

After Lysenko's claim at our lunch table, I sat silently for a while, wondering what I should do next. Should I pass over his self-justifying claim, or should I challenge him? Finally, I realized that I had in my hands the chance of a lifetime. Never again would I have an opportunity to test the most infamous scientist of the twentieth century. If I made him angry and he blurted out something revealing, maybe other researchers of Lysenko would learn something useful. I hoped that the fact that he had been discredited in the Soviet Union meant that he would not be able to bring the wrath of the secret police down on me, as he had on his previous critics. (Actually, after this conversation until Lysenko's death in 1976 I was declared persona non grata by Soviet authorities, but I am not certain that Lysenko had anything to do with the change in my status.)

I decided that I would refute his claim but do so in the calmest, most academic way possible, based on my recent research. I would use the example of his most famous adversary, Nikolai Vavilov, a world-renowned geneticist and creator of the world's largest collection of plant seeds, who, because of Lysenko, ultimately died of starvation in detention.[1]

Lysenko had many facilitators in his quest to rid himself of his opponents. Their link to Lysenko was not so much to his biology as to a shared antipathy to bourgeois specialists and anyone else who tried to exert superior knowledge. It began in the 1920s, and by the early 1930s Lysenko's critic Vavilov had already encountered strong resistance in the party organization and among the postgraduate students at his own Institute of Plant Breeding.[2] These were often veterans of the bitter Russian Civil War who had been radicalized and taught to distrust, even despise, people like Vavilov, who came from a privileged background and was educated, before the revolution, in laboratories in England, the United States, and other countries. Lysenko was more attractive to them because of his peasant origins than because of his biological views. And he further enticed them with his wild promises about the future of Soviet agriculture. These radical students and activists disputed the leadership of Vavilov, whom they saw as a representative of the old order. As Russian historian of science Eduard Kolchinsky observed, "By 1932 Vavilov had already lost his independence in personnel management and control over institutions he administered."[3]

As the years progressed, Vavilov was gradually overwhelmed, and Lysenko was catapulted into power. An honest man, Vavilov could never make promises equal to those of Lysenko, although he made several pathetic attempts. Vavilov knew his own vulnerabilities and vainly hoped to satisfy Lysenko by creating a division of function between them. Lysenko would be a great practical plant breeder, a green thumb, while Vavilov and his geneticist colleagues would be the theoreticians. But Lysenko was too ambitious to accept this modest role, and helped by others eager to assume the resulting vacant positions, he destroyed Vavilov and his colleagues.

Now, sitting beside Lysenko decades later, I started out by saying:

You know, I know that you are correct in saying that you were never a member of the Communist Party. But you frequently criticized Vavilov and other Russian scientists in ways that were certain to attract the attention of the secret police. For example, at meetings in 1935 and 1939 at which Stalin was present you said that there were saboteurs both in Soviet industry and in Soviet agriculture, and you named Vavilov as one of those traitors. You also stated that you were just a simple agronomist, not a Communist Party member, not a politician. And Stalin broke out with "Bravo, Comrade Lysenko, Bravo!" Yet I know that, far from being a traitor, Vavilov was dedicated to the Soviet cause and did everything he could to improve Soviet agriculture. But Vavilov recognized the importance of modern genetics in that effort, which you opposed. So you denounced him in the presence of Stalin, won Stalin's approval, and the secret police did the rest. Vavilov, as you know, died in their custody.[4]

Lysenko abruptly stood and left the table. I sat alone, eating my soup. After about ten minutes, to my astonishment, Lysenko returned and sat down beside me, saying:

You are mistaken in your understanding of me. You think that I am a part of the Soviet oppressive system. But I have always been an outsider. I came from a simple peasant family, and in my professional development I soon encountered the prejudices of the upper classes. Vavilov came from a wealthy

Russian geneticist Nikolai Vavilov, one of the victims of Lysenkoism. (RIA Novosti/Science Source)

family, was as a consequence well-educated, and knew many
foreign languages. When I was a boy I walked barefoot in
the fields and I never had the advantage of a proper education. Most of the prominent geneticists of the 1920s and 1930s
were like Vavilov. They did not want to make room for a
simple peasant like me. I had to fight to be recognized. My
knowledge came from working in the fields. Their knowledge came from books and laboratories, and was often mistaken. And once again, I am now an outsider. Why do you
think I was sitting alone here at this table when you came
up? No one will sit with me. All the other scientists have ostracized me.[5]

I knew he was correct in describing Vavilov's privileged background. But the words that most struck me were, "You think that
I am a part of the oppressive Soviet system." Yes, I did think that. Was
it possible that he was not lying—that he actually thought he was
somehow outside that system? In the beginning, as a humble farmer
trying to make his way, he may have considered his "outsider"
status as natural. But he became a major symbol and a stalwart of
the Soviet system, benefitting mightily from it, sacrificing his colleagues in the process, and totally implicating himself in the Soviet
regime.

And then Lysenko said something that amazed me. At this time,
in the 1970s, there were several thousand Jewish refuseniks in Moscow,
many of them scientists and some of them geneticists, who had
applied to emigrate to Israel, been refused, and then were fired from
their jobs. They lived hand-to-mouth, often supported by friends and
relatives in the West who managed to get money and food to them
(I participated in these aid efforts).[6] "I sympathize with the Jewish

refuseniks," Lysenko said. "Many of them are scientists who have been ostracized by the Soviet establishment because they applied to emigrate to Israel. Now they have no jobs and no place to turn. They are alone like me."[7]

Lysenko was obviously attempting to win my sympathy. Doubtless, he knew I would write about this meeting. But even though Lysenko had lost his scientific authority, he was still a member of the Academy of Sciences, with a good salary, an office, and many privileges, including access to special grocery and clothing stores.[8] Indeed, the luxurious House of Scientists in which we sat, with its excellent food at very reasonable prices, was a perquisite of which the refuseniks could not dream. Lysenko's attempt to compare his position to that of the Jewish refuseniks was grotesque.

Nonetheless, in his self-serving description of himself I recognized a certain truth about Soviet history: the savage consequences of class hatred when linked to state power. Without state power to support him, Lysenko would have remained a simple agronomist, preaching his special approach, gaining little attention from the academic establishment, and doing no physical damage to anyone. With state power behind him, he caused a great tragedy.

But one should see the human elements here as well. Lysenko undoubtedly saw himself as a peasant fighting aristocrats. He also, at least originally, probably believed in his simple agricultural nostrums. Later, when in decline, he resorted to dishonest methods in his own research, concealing his failures.[9] He thought that what counted in obtaining more milk from cows was not their genetic constitution but the care they received (my grandfather on his Indiana farm thought the same thing). Lysenko took very good care of his cows, fed them copiously, and even made sure their stalls were clean. He was certain they would return his favors by giving lots of milk. He could not understand why purebred cows, some imported at great cost to Russia from the British islands of Guernsey and Jersey, where they were originally bred, should give more milk just because they had advantaged ancestors. Similarly, he could not understand why Vavilov and the privileged scientists he fought against should be better scientists than he. When Lysenko realized that the Soviet system gave him a formidable weapon against his enemies, he eagerly seized the chance. The simple peasant, nursing grievances against those he viewed and resented as his "social betters," became a tyrant who sent dozens of people to their deaths. The leaders of the Soviet Union of his time—Stalin and Khrushchev—knew little about modern genetics and could not see the errors of Lysenko's scientific views. They saw only that Lysenko praised them and their rule. Both had humble backgrounds like Lysenko and criticized the privileged Western world. After this conversation with Lysenko, I did not change my view about his personal responsibility for the tragedy of Soviet genetics, but somehow I better understood the motives behind his tyranny.

Almost twenty-five years after that 1971 meeting, after Lysenko's death in 1976 and the fall of the Soviet Union in 1991, I found myself back in the House of Scientists dining room, where the luxury was looking a bit tired but still impressive. I was there with George Soros, the wealthy American philanthropist helping Russian science at a moment of great need. Soros sympathized with the geneticists who had suffered under Lysenko's rule. Some had been imprisoned for decades. A few survived, were released from prison, and were now living freely but poorly in Russia. Soros had suggested a banquet for these geneticists, which is why we were there that day.

Accompanying Soros and me at the banquet was Valery Soyfer, a geneticist who had written a history of Lysenkoism while living in the Soviet Union and then emigrated to the United States.[10] As we sat at this poignant reunion of persecuted geneticists, I stared over at the corner table where, many years before, I had heard Lysenko describe these scientists as aristocratic traitors to the Soviet cause. They certainly did not look like aristocrats. Many wore frayed clothing, and they were bent over from their sufferings in the labor camps. Soros asked all of the aged geneticists to tell stories of their colleagues. Sergei Chetverikov, a pioneer in the development of the "biological synthesis" of the 1920s, was arrested and sent into exile and never returned to his main research topic. Theodosius Dobzhansky fled to the United States to escape political controls and became a famous scientist there. Georgi Karpechenko, the first person to create a new species through polyploidy speciation, was sentenced to death and executed in 1941. Nikolai Kol'tsov, the early leader of geneticists in the 1920s, was accused of ideological sins, dismissed from his position, and left the field. Nikolai Vavilov, arrested in 1940, starved to death in jail in 1943. Nikolai Dubinin, a prominent geneticist both before and after Lysenko, abandoned the field in 1948 and worked for many years as an ornithologist, returning to his main studies only after 1965. D. D. Romashov was arrested twice but released because of illness; his wife died in prison. N. V. Timofeev-Resovskii, eminent geneticist, emigrated to Germany, was arrested in Berlin, and returned to Russia only many years later. All in all, several hundred geneticists were repressed.

We should recognize that we cannot be certain that these scientists were all arrested because of their views on genetics. People all over the Soviet Union were arrested in those years for a variety of alleged crimes, usually falsely. But many Russian geneticists believed they were arrested because they refused to accept Lysenko's doctrines—and in many cases they were probably correct.

Lysenkoism forced some Russian scientists out of genetics. Nikolai Dubinin, a prominent geneticist, abandoned the field in 1948, returning only after 1965. (RIA Novosti/Science Source)

The task of the scholar evaluating Lysenko is complicated: on the one hand, Lysenko was responsible for the deaths of many of his colleagues; on the other hand, geneticists in recent years have learned that acquired characteristics *can,* in at least some instances, be inherited, one of Lysenko's central claims. Was Lysenko right in part? Was he right for the wrong reasons? To answer these questions, we must look in more depth at Lysenko's scientific work.

6 | LYSENKO'S BIOLOGICAL VIEWS

> No positive results can be obtained from work conducted from the
> standpoint of Lamarckism.   —TROFIM LYSENKO

EVERY PERSON DESERVES his day in court. Let us now look at Lysenko's biological views, evaluating them as fairly as possible. Trofim Denisovich Lysenko was born in 1898 in Ukraine, near the city of Poltava, where he grew up in a peasant family. He received an education as a practical agronomist at the Horticultural Institute of Poltava, continued his studies and research at several different locations in Ukraine, and after 1925 began to investigate the vegetative periods of agricultural plants at the Gandzha Plant-Breeding Station in Azerbaijan.

Between 1923 and 1965, Lysenko published approximately four hundred different works, all of which I have read.[1] Many are duplications, almost word-for-word, of earlier publications. By 1948 he had developed all the major components of his biological "system."

Lysenko's views on biological development were contained in a vague doctrine, the "theory of nutrients." He used the word "nutrient" *(pishcha)* in a very broad sense, including, for plants, environmental conditions such as sunlight, temperature, humidity, chemical elements in the soil, and gases in the atmosphere. To Lysenko, any approach to the problem of heredity must start by considering the relationship between an organism and its environment, and the environment, in the final analysis, determines heredity. Although he insisted, in answer to critics, that he did not deny the existence of genes, he paid no attention to them. He believed that the whole cell—not some constituents within it—was the carrier of heredity.

Those with a weak knowledge of the history of heredity studies may see in Lysenko's belief a prescient insight into cytoplasmic inheritance (the non-Mendelian transmission of genes that occur outside the nucleus). However, there was nothing new here. This viewpoint was held by dozens of cytologists preceding him, including C. Benda (1901) and F. Meves (1908) in Germany, J. Duesberg (1913) in Belgium, and E. Fauré-Fremont (1908) and A. Guillermond (1913) in France.[2] Lysenko, unenlightened by the great advances made in genetics in the first decades of the twentieth century, particularly in the United States and Great Britain, was following an old tradition.

Lysenko was a firm defender of the doctrine of the inheritance of acquired characteristics. The "internalization of environmental conditions," which he considered the means by which the heredity of any type of organism is acquired, is obviously a type of such inheritance. And Lysenko himself stated his position unequivocally, tying such inheritance to the materialistic worldview of Marxism (the theory that matter—physical, biological, and social—is the only reality). Just why classical genetics, based on DNA and genes, was any less materialistic did not concern him: "A materialistic theory of the development of living nature is unthinkable without a recognition of the necessity of the inheritance of individual differences by the organism in definite conditions of its life; it is unthinkable without a recognition of the inheritance of acquired characteristics."[3]

The most important influence on Lysenko's theory of nutrients was his work at the end of the 1920s and the beginning of the 1930s on the effects of temperature on plants. Lysenko concluded that the relationship between an organism and its environment could be divided into separate phases or periods, during which the requirements of the organism differed sharply. He sometimes labeled his views "the theory of phasic development of plants," although the "theory of nutrients" was a more comprehensive title, describing both plants and animals.

As an example, Lysenko pointed to the fact that many cereals, including both spring and winter varieties of wheat (winter wheat is planted in the fall and ripens the following spring or summer), require lower temperatures near the beginning of their growth periods than at the end. Whether the grain is normally planted in the fall or in the spring, the period immediately before and after germination is much colder than the period during maturity. This relative coldness, he said, is not only a circumstance of the climate but is, at the present point in the evolution of many of these cereals, a necessity for their full life cycle.

Using this approach, Lysenko maintained that he could convert a winter wheat variety into a spring variety. His best-known example was the case of the *Kooperatorka* winter wheat, which he called in 1937 "our most prolonged experiment at the present time."[4] On March 3, 1935, Lysenko sowed this variety of winter wheat in a greenhouse that was kept until the end of April at a very cool temperature, using snow as a coolant. After the vernalization treatment (the process of hastening the flowering of plants by exposing them to low temperatures), the snow was removed, and the temperature was raised. Originally, there were (only!) two *Kooperatorka* plants, but one perished, Lysenko said, as a result of pests. The sole surviving plant eared on September 9, proving to Lysenko that vernalization had worked, since *Kooperatorka* normally matures in the spring. Grain then taken from the plant was immediately sown, again in a greenhouse, where it eared as an F_2 generation (an F_1 generation is the offspring of the cross of the first set of parents, while the F_2 generation is the result of a cross between two F_1 individuals) at the end of January. On

Lysenko examining wheat plants. (Sovfoto/Getty Images)

March 28, 1936, the third generation was sown, producing seed in August 1936. Thereafter, the grain acted as a spring variety, and Lysenko maintained that its "habit" had been converted.

All that can be concluded from such an experiment is that Lysenko's methods were incredibly lacking in rigor. The fallacy of basing scientific conclusions on a sample of two need not be emphasized. The *Kooperatorka* was probably heterozygous (having two different alleles, one dominant and one recessive, for a single trait); the one surviving plant could well have been an aberrant form. Even had several plants survived, a selection out of the variations would naturally occur. If one attempts to convert a winter wheat into a spring wheat and sows in the spring, one will be able to gather in the fall only the grains from the plants that have matured. The effects of selection could be avoided, or rather determined, by using a variety of known purity coupled with careful statistical studies of many plants over a number of generations, including statistics for plants that did not mature and large control groups of nonvernalized plants. But Lysenko hated statistics, had no concept of rigorous control samples, and said that in a time of urgent agricultural need one should not delay progress with detailed studies. He ignored the fact that attempts to duplicate his results were made outside the Soviet Union and failed.[5]

Lysenko believed that the vernalization phase applied not only to cereals but to all plants. He believed that for many, the vernalization phase was the first of several stages. Cotton, for example, needs very warm temperatures in its first stage and relatively lower ones in its last, or boll-ripening, period. Even though the cotton plant's requirements are the opposite of wheat's, Lysenko wrote, the first stage of the cotton plant can also be called its vernalization phase, a period in which temperature is fundamentally important. Unless a plant has passed through this vernalization phase, it cannot reach maturity and bear fruit.

From this observation Lysenko proceeded to a generalization:

The changes in the environmental conditions which developing plants need indicate that the development of annual seed plants from germination of the seed to the ripening of the new seeds is itself not uniform in type, not uniform in quality. The development of plants consists of separate qualitatively different stages, or phases. To pass through the different phases of development, the plants require different external conditions (different nutriment, light, temperature, etc.) Phases are definite necessary stages in the plant's development, and serve as the basis of the development of all the plant's particular forms—its organs and characters. The different organs and characters can develop only at definite phases.[6]

Lysenko's major errors were not in his subject of study but in his methods and conclusions. The study of the phasic development of plants is a perfectly legitimate topic in plant science; it was an established subject in Western biology and even had a name, "phenology." It goes back to Linnaeus in the eighteenth century; by the mid-nineteenth century became a full-blown branch of biology, especially in Germany and England; and continued into the twentieth century, with its own organizations and publications.[7] Eventually, it involved thousands of researchers. A vast literature exists on the cold treatment of plants, reaching back centuries before Lysenko's work.[8] In 1662 John Evelyn gave a report to the Royal Society in London describing how he could encourage seeds to germinate in the spring by artificially exposing them to cold-moist conditions between layers of soil or peat.[9] The technique was known in the United States as early as 1854 and was also the subject of research in Germany by G. Gassner before the end of World War I.[10] What Lysenko called "vernalization" was often called elsewhere "cold stratification." The fact that seeds of various kinds require certain conditioning periods, during which temperature and moisture are critical factors, is commonplace knowledge in plant propagation.

More importantly, vernalization is explainable and best understood when based on the classical genetics that Lysenko rejected. During the years Lysenko was in power, geneticists and plant physiologists elsewhere were researching the cold requirement of vernalizable plants, trying to identify the genes responsible. By the time of Lysenko's fall from power, they had identified up to six genes.[11] In some

John Evelyn gave a report in 1662 to the Royal Society in London describing what Lysenko later called "vernalization." (Middle Temple Library/ SPL/Science Source)

cases, including wheat and rye, the cold requirement is caused by recessive alleles; in others, including barley and some varieties of wheat, it is caused by dominant alleles. Lysenko denied the very existence of dominant and recessive alleles.[12] This research today has reached an advanced stage, promising ways to trigger early germination using procedures far superior to Lysenko's crude methods.[13]

The actual processes that take place within seeds before germination are extremely complex, involving biochemical and physical changes, including natural inhibitors and hormone balances. In an effort to manipulate these processes, researchers have not only controlled the temperature and humidity of the seeds but have alternated such changes in complex patterns; scraped the seeds with sandpaper; and treated them with various chemicals, including acid solutions, to render the seed coat (testa) more permeable.

Not every potentially useful technique that works under laboratory conditions can be economically employed. Spreading seeds on the ground or in trays, applying water or chemicals at controlled temperatures for possibly weeks, or requiring special buildings with heating or cooling facilities all require considerable capital expenditure and enormous labor. Furthermore, the process of vernalization provides an ideal scenario for the spread of certain fungi and plant diseases. Many researchers have concluded that vernalization often involves greater losses than gains.

In the Soviet Union of Lysenko's time, where electricity and refrigerating equipment were often lacking, it proved nearly impossible to keep the seeds in uniform conditions over long periods. Sometimes, the seeds became too hot, too cold, too wet, or too dry. Some seeds germinated too rapidly, some too slowly, and some not at all. But perhaps these very losses also provided excuses: if vernalization on a particular farm failed, this could easily be blamed on the conditions, not the process.

Lysenko used the term "vernalization" in an exceedingly loose way, covering almost anything done to seeds or tubers before planting. The vernalization of potatoes that Lysenko promoted included inducing the tubers to sprout before planting. Gardeners all over the world have followed this practice for centuries, cutting potatoes into sections with at least one eye and then soaking these in water before planting.

Many experiments with vernalization worked both ways. Vernalization was only rarely used to make the previously impossible possible: growing crops in a region they had never been grown in before because of the climate. Rather, it was usually directed toward making traditional crops ripen earlier or growing a grain that because of the length of its growing season was difficult to successfully harvest before the frost. In these kinds of experiments, the evidence can be manipulated very easily, and sloppiness in record keeping and ignorance of statistical methods can conceal results from an honest researcher. A two- or three-day difference in the date of the ripening of a grain is a very inconsiderable period, subject to many different kinds of interpretation. Enthusiasm in claiming victories for vernalization (typical in Soviet media) would go a long way in conditions of inaccurate records, uneven controls, variable agronomic conditions, impatience about verification, willingness to discount contradictory evidence, and impure plant varieties.

Lysenko believed that the vernalization phase is only one of several different stages that plants must pass through to bear fruit.[14] However, he never fully described just what these other stages are. He mentioned that for many cereals the "photo phase," during which the duration of daylight becomes critical, immediately succeeds vernalization. While in each of the two phases one factor (temperature or light) becomes critical to the development of the organism, he emphasized that they alone do not guarantee correct development.

Lysenko tried to overcome his difficulty in defining his phases by maintaining that under normal conditions all the factors except the critical one are present in correct measure to enable the organism to successfully pass through the phase.

Lysenko claimed that he had developed a general law of the development of plants, the "law of phasic development." A priori there is, of course, nothing wrong with an attempt to develop such a general phasic law; its success depends on its rigor, usefulness, and universality. The implication of Lysenko's use of the term "normal" is that his critical factors might not be critical in other districts, under other conditions. What did "normal" mean: The conditions of cold Russia, temperate Spain, or tropical Brazil? It is not surprising that Lysenko, in a northern region, might select temperature as critical for successful early planting. But temperature might not be as important as, for example, water availability or soil conditions in southern areas of the world. Can any one factor be "critical" everywhere? And if not, how can any law of the type Lysenko was striving for be truly "universal"? We are evidently left with the correct but unoriginal conclusion that in northern regions temperature is very important for plants, in semidesert regions water is crucial, and in dark or cloudy regions light is a key variable—views not only common throughout the world, but close to common sense.

By 1935 Lysenko was reaching beyond simple studies of vernalization to a general theory of heredity. He complained that classical geneticists could not predict which characters would be dominant in hybridization and worked primarily by making many thousands of combinations. Lysenko's impatience—linked with pressure from the Soviet government in its quest for rapid economic expansion—drove him to the hope for shortcuts. He believed that dominance depended on environmental conditions: "We maintain that in all cases when a hybrid plant is given really different conditions of existence for its development this causes corresponding changes in dominance; the dominant character will be the one that has more favorable conditions for adapting itself to its development."[15] This assertion simply denies the basic principles of Mendelism and has not been substantiated.

Lysenko also denied the distinction between phenotype (the observable characteristics of an organism that result from its genotype and the environment) and genotype (the genetic makeup of a cell), even over the distance of one generation.[16] He maintained that "all the properties, including heredity, the nature, of an organism, arise de novo to the same degree to which the body of that organism (for example, a plant) is built de novo in the new generation."[17] The obliteration of this separation lay at the bottom of much of Lysenko's writings and repudiates the whole structure of modern genetics.

Lysenko defined heredity as "the property of a living body to require definite conditions for its life, its development and to react definitely to various conditions."[18] According to Lysenko the heredity of a living body was built up from the conditions of the external environment over many generations. Each alteration of these conditions led to a change of heredity, a process he called the "assimilation of external conditions." Once assimilated, these conditions become internal—that is, a part of the nature, or heredity, of the organism: "The external conditions, being included within, assimilated by the living body . . . become particles (chastitsy) of the living body, and for their growth and development they in turn demand that food and those conditions of the external environment, such as they were themselves in the past."[19] In the last part of this sentence, Lysenko referred to the part of his biological system that avoided a totally arbitrary plasticity of organisms. The mechanics of the transition from "external conditions" (temperature, moisture, nutriments, and so on) to "internal particles" was, to say the least, unclear, but

Lysenko did achieve a concept of material carriers of heredity. These internal particles may seem at first glance to be genes, but it is clear from Lysenko's description and later comments that they are not. Rather than being unchanging, or relatively unchanging, hereditary factors passed from ancestors to progeny, they are internalized environmental conditions subject to easy change—not just in their expression but also in their most basic structure.

Lysenko's particles did provide a rough sense of heredity in the simplest and most customary sense. He said that if an organism exists in external surroundings similar to those of its parents, it will then display characters similar to its parents'. If the organism is placed in an environment different from that of its ancestors, its course of development will differ from theirs. Assuming the organism manages to survive, it will be forced, Lysenko thought, to assimilate the external conditions of its new environment. This assimilation leads to a different heredity, which in several generations may again become "fixed" as it was in the early environment. In the intermediate, or transition period, Lysenko believed that the heredity of the organism is "shattered" and therefore unusually plastic. During this period Lysenko believed it is possible to transform that organism and even create a new species. He claimed that he and his colleagues had accomplished that in a number of instances; one of the best-known being the alleged transformation of a hornbeam tree into a hazelnut.[20] None of these claims has been substantiated.

According to Lysenko, one could shatter the hereditary stability of an organism in three different ways. One could place the organism in different external environments, as already described. This method, he thought, was much more effective at certain stages (for instance, vernalization) of the development process than others. Or, one could graft a variety of plant onto another, thereby "liquidating the conservatism" of both stock and scion. Or finally, one could cross forms differing markedly in habitat or origin. Each of these methods was attempted in Lysenko's experiments.

Lysenko's particles remind one a bit of Charles Darwin's "gemmules," which were supposedly given off by every cell or unit of the body. One might add that in Darwin's time, without any knowledge of genes, the theory explained phenomena that otherwise could not be explained. Darwin, furthermore, was aware of the speculative character of his gemmules and labeled them "provisional." Lysenko inadequately and incorrectly accounted for phenomena that were much better explained by existing biological science. Darwin's effort was innovative and useful, even if later replaced by a better explanation, while Lysenko's retrogressive theory ignored a mountain of achievements in the science of heredity since Darwin's time.

Lysenko's view of the possible types of inheritance included the case of particulate, or mutually exclusive, inheritance, but went far beyond it. He borrowed his system largely from Timiriazev (1843–1920), who in turn had been influenced by earlier biologists. Here again, Timiriazev's scheme originated in the late nineteenth and early twentieth centuries, making it seem fairly plausible. By the time Lysenko espoused it, genetics had created a far superior scheme, which Lysenko never mastered. Timiriazev's and Lysenko's categories of inheritance can best be described in terms of a diagram given in Hudson and Richen's careful study;[21] Lysenko described the same scheme in his *Heredity and Its Variability*.[22]

Examples of simple inheritance, in which only one parent is involved, would include all types of asexual and vegetative reproduction (self-pollination in plants such as wheat, propagation from tubers or cuttings, and so forth) and parthenogenesis. Complex inheritance involves two parents, and according to Lysenko, this "double heredity gives rise to a greater viability of the organisms,

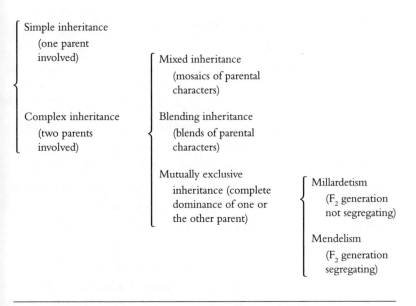

Lysenko's classification of the types of inheritance.

and to their greater adaptation to varying living conditions."[23] Lysenko therefore thought that the offspring of two parents possessed, in potential, all the characters of both parents, and he looked with disfavor upon inbreeding or self-fertilization, thinking it led to the narrowing of an organism's potentialities.[24] For somewhat less clear reasons, he opposed artificial insemination. In the case of double heredity with unrelated parents, the characters that would actually be displayed depended on first, according to Lysenko, the environment in which the organism was placed and second, the unique properties of the organism. The interaction of the environment and these unique properties led to "types" of complex inheritance: mixed, blending, and mutually exclusive.

To Lysenko, progeny with "mixed" heredity displayed clear (unblended) characters of both parents in different parts of their bodies;

examples would be variegated flowers, piebald animals, and grafts of the type known to geneticists as "chimeras" (mosaic patterns of genetically distinct cells formed by artificially grafting two different plants). Lysenko's best-known example of mixed heredity was the supposed graft hybridization of tomato plants by Avakian and Iastreb, in which the stock reportedly influenced the coloration of the fruit of the scion. This experiment was investigated by Hudson and Richens, who concluded that it was of doubtful validity.[25] If the tomato plants were heterozygous, and if stray cross-pollination occurred, the results could be explained in terms of standard genetics. Lysenko was a strong proponent of graft hybridization and did some interesting and disputed work in the field. The topic was not new; Darwin believed in the possibility of true graft hybrids, as did many nineteenth-century biologists and selectionists. Luther Burbank, who like Lysenko kept very poor records, advanced a tentative theory of sap hybridization similar to graft hybridization. Michurin's "mentor theory" postulated the influence of the stock by the scion. The field remains controversial, but Lysenko's failure to use proper experimental controls means that he was not a reliable participant in the debate.[26]

"Blending" inheritance was, to Lysenko, the merging of characters in the hybrid in such a way that they are intermediate between those of the parents. Many cases of such inheritance are known. It is obvious, for example, that the children of parents of distinctly different skin color are frequently intermediate in color, and a whole spectrum of intermediate forms may occur with no clear relationship to the Mendelian ratios. Lysenko's and modern geneticists' interpretations differ in that the latter see continuous variation as the result of a series of independent genes that are cumulative in effect but still function discretely, while Lysenko spoke simply in terms of blending.

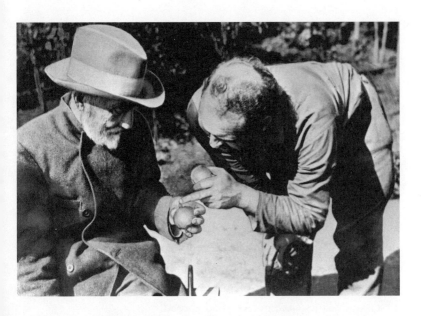

In the Stalinist period, botanist Ivan Michurin was posthumously heralded as a leader in the theory of evolution. (RIA Novosti/Science Source)

"Mutually exclusive" inheritance was the term Lysenko used to cover the phenomenon of complete dominance. Lysenko did not see dominance in the customary terms of the mechanism of allelic pairs, only one of which in the hybrid form is expressed in the phenotype, but instead in terms of the relationship of the organism to the environment. He believed there were no dominant and recessive genes, only "concealed internal potentialities" that may or may not "find the conditions necessary for their development." Those characters would be expressed that found the "proper conditions."[27] Lysenko felt that this theory provided a better means for manipulating heredity, since he believed that the inevitable dominance of one character over another did not exist, and characters could be altered by man.

Lysenko saw two different types of mutually exclusive inheritance, which he called "Millardetism" and "Mendelism," or "so-called Mendelism." Millardetism, named after the French botanist, described hybrids that in subsequent generations supposedly never display "segregation" (the separation of allelic genes into different gametes). The dominance displayed in the F_1 generation continues, Lysenko reported, in all other generations. Lysenko maintained that this was not surprising since his general theory of the expression of characters rested on the relationship of the organism to the environment; therefore, the correct environment would always cause the appearance of the appropriate character. Lysenko's followers cited a number of experiments that allegedly supported this. Nothing in classical genetics explains these particular cases, although it is not difficult to imagine errors that might lead one to such a conclusion.[28] Lysenko's results were not verified abroad.

"So-called Mendelism," the last of Lysenko's types of inheritance, refers to hybrids that do segregate in F_2 and subsequent generations. Lysenko considered them isolated cases and insisted, like Timiriazev, that Mendel did not actually discover this type of inheritance. Lysenko considered the Mendelian laws to be "scholastic" and "barren." They did not reflect the importance of the environment and did not permit the prediction of the appearance of characters before making empirical tests for each type of organism.

Many in the West have speculated that the reason Lysenko's views became dominant in Soviet biology was because of their promise to improve human beings and create "A New Soviet Man."[29] If Soviet leaders believed that characteristics acquired during a person's lifetime could be inherited, they thought, then they would believe that a unique Soviet individual would emerge all the more quickly.[30] This speculation is historically false. Absolutely no evidence for this viewpoint can be found in Lysenko's writings or in the speeches of Soviet leaders during his heyday. Influenced by the debates of the 1920s and the Communist consensus that emerged, Lysenko confined himself at first to plants and a little later, to animals. He criticized all efforts to apply biology to humans. By the time Lysenko became powerful, in the mid- to late 1930s, the eugenicist and racist views of Nazi Germany had thoroughly discredited efforts to explain the emergence of superior individuals on the basis of biological theories. In 1958 (when Lysenko was still firmly in power), *Pravda* spoke of N. K. Kol'tsov (1872–1940), the prominent Soviet biologist associated with eugenicist views in the 1920s, as a "shameful reactionary who is known for his wild theory that preaches 'the improvement of human nature.' "[31] Yet "the improvement of human nature" is precisely the reason given by some writers for the entire Lysenko affair.

Concluding this chapter on Lysenko's biological views, we see that Lysenko made many claims that modern genetics did not accept in his time and still does not accept today. In his denial of Mendel's laws, in his refusal to accept the distinction between phenotype and genotype, in his claims that he could create new species, and in his definition of heredity as "the property of a living body to require definite conditions for its life," he departed from both the science of his time and contemporary biology. Today, even though our knowledge of genetics has greatly improved since Lysenko's heyday, there seems no reason to accept his claims.

As for the concept of the inheritance of acquired characteristics, Lysenko had little ownership of that doctrine. Belief in the inheritance of acquired characteristics "solaced most of the biologists of the nineteenth century," as a prominent geneticist of the twentieth century observed.[32] Thus, Lysenko could cite Darwin as well as Timiriazev and Michurin in support of the view.[33] One might add, parenthetically, that the surprising aspect of Darwin's attitude toward the inheritance of acquired characteristics was not that he believed in it (which he did) but that he relied so little on it for his great theory.

Whether Lysenko was a Lamarckist is an interesting question, despite the fact that he is usually considered to be one. Lysenko himself denied that he was a Lamarckist. In fact, he said "no positive results can be obtained from work conducted from the standpoint of Lamarckism."[34] And if the eighteenth-century savant Jean-Baptiste Lamarck could somehow be resurrected, he would probably agree that Lysenko was not a Lamarckist. There is in Lamarck no equivalent to Lysenko's theory of "shattering heredity" or his theory that heredity is a metabolic process. Furthermore, Lamarck emphasized "use and disuse" and the "trend toward complexity," neither of which played a role in Lysenko's thought. Both systems contain the principle of the inheritance of acquired characteristics, but so do dozens of other past biological systems. So once again, usage triumphs over accuracy, and Lysenko is probably indelibly identified as a "Lamarckist."

As we have seen, Lysenko's scientific work was shoddy and unsubstantiated. But his association with the doctrine of the inheritance of acquired characteristics leads some to claim that epigenetics validates some of his ideas. To determine if this is the case, we must look at this new science in more detail.

7 | EPIGENETICS

The epigenetics revolution is underway.   —NESSA CAREY

IN THE LAST TWENTY YEARS, a major shift has occurred in our knowledge about inheritance. On the question of the inheritance of acquired characteristics, the shift is particularly dramatic. This theory, discredited during much of the last century, is now accepted by many biologists as valid, at least in some instances and at certain times. However, some prefer not to use the term, speaking instead of "epigenetic transgenerational inheritance." According to such inheritance, characteristics acquired during the lifetime of an organism can be inherited.[1] August Weismann, William Bateson, Thomas Hunt Morgan, and other pioneers of classical genetics would have been staggered to hear this news.

The classical view of genetic inheritance that developed in the twentieth century as chromosomes, genes, and DNA gradually came to be understood was the following: DNA is like a template that molds messenger RNA, which in turn creates proteins that make up the body of the organism. The passage of information comes entirely from the DNA in an outward direction, never in an inward direction. The DNA of genes is the master director and does not change, except for rare mutations or as a result of extreme influences, such as radiation. People speak of "selfish genes" merely wanting to replicate themselves, uninfluenced by environmental factors.[2]

Long before epigenetics became a subject of great attention, the classical model of genetics just described began to be modified. The following were among the most important developments. In the 1950s

Joshua Lederberg developed the concept of plasmids, which he described as "any extrachromosomal hereditary element."[3] In the 1950s and the 1960s, Barbara McClintock published seminal articles that described "mobile controlling elements," or "jumping genes" (genes that landed randomly), that regulated specific gene expression without changing gene structure.[4] Both Lederberg and McClintock received Nobel Prizes. In 1958 D. Nanney published an important article titled "Epigenetic Control Systems," in which he spoke of "dynamic inheritance."[5] In 1965 François Jacob, André Lwoff, and Jacques Monod received a Nobel Prize for their explanation of gene expression, that is, whether a gene is "on" or "off."[6] In 1991 O. E. Landman published an article expressing incredulity that so many biologists remained convinced of a basic contradiction between the principle of the inheritance of acquired characteristics and molecular biology. On the contrary, he wrote, the inheritance of acquired characteristics is "fully compatible with current concepts of molecular genetics," and he pointed to examples of such inheritance.[7] These developments provided an important background and a prehistory to the blossoming of epigenetics in the early years of the twenty-first century.[8]

With the recent development of epigenetics came a rethinking of basic issues of inheritance, including the concept of the inheritance of acquired characteristics.[9] In the new interpretation, DNA and genes are still fundamentally important (something supporters of Lysenko's views often neglect), but they are seen less and less as "the directors" of inheritance and more and more as its informational depository, storing individual packets of information that are sometimes used and sometimes not, according to other influences, including environmental ones. In the new view, DNA might be compared to a giant library, containing thousands of books (humans have about twenty thousand genes). Some of these books are read often, some rarely, and some maybe even not at all under normal

The team of François Jacob, Jacques Monod, and André Lwoff won the 1965 Nobel Prize for work on the regulation of transcription of DNA sequences. (Jean-Claude Deutsch/Getty Images)

conditions. Various kinds of controls can be placed on certain books so that procuring and reading them can be quite difficult. Some books are "featured" and easily obtained, while others are prohibited, at least temporarily. And the books in the library are regularly being supplemented, changed, and even discarded. As this relates to genetics, some biologists began to speak of "genome inconstancy."[10] The director of all these processes is less clear than under the earlier simplified view; there seem to be multiple directors, and environmental factors are among them. Retaining the metaphor of the library just a bit longer, the books in the library that can be read are comparable to genes that are expressed, or turned on; the books that cannot be read are akin to genes that are not expressed, or turned off.

What mechanisms in the DNA itself account for these actions? Chemical groups, it turns out, can attach themselves to DNA in a way that does not change the DNA itself but does change its expression. One of the most important, according to epigenetics, is a methyl group (CH_3), three hydrogen atoms attached to a carbon atom. Such methyl groups can attach themselves to DNA and inhibit or prevent the expression of genes. Furthermore, it appears that such "methylated DNA" in some cases (how many is a subject of dispute) can persist not only during the life of the cell but also in the cells descended from the original methylated one. In such a descent line, the genes that have been turned off stay turned off, at least for some generations. Environmental influences can cause methylation, and they can also cause "demethylation." In other words, epigenetic inheritance seems to be a reversible process. And it appears that other chemical groups and the physical configuration of DNA—the way it is folded—can have similar effects.

Such epigenetic inheritance is well established in plants, less so in mammals. As two specialists recently wrote in the leading biological journal *Cell,* "In plants, evidence for heritable epigenetic variation is more than a half a century old."[11] More controversial is the hypothesis of such transgenerational inheritance in mammals, including humans. In mammals the methylation of DNA is often eliminated at the time of reproduction. Within the same article, the specialists maintained that epigenetic inheritance is not proved in mammals.[12] But while controversy remains, hundreds of publications postulating such epigenetic transgenerational inheritance in humans and other mammals have appeared in established academic journals and in the popular press. Unfortunately, many of the reports in the popular press make exaggerated claims. In 2010 the German newsmagazine *Der Spiegel* headlined the new developments as a "Victory over Genes."[13] On August 15, 2014, two authors published an article in

Science in which they stated, "Human and animal studies have demonstrated that the prenatal environment affects adult health and disease . . . adverse metabolic health outcomes can be transmitted multigenerationally."[14] However, most studies in mammals indicate that the effects do not persist more than two or three generations.

A well-known example of epigenetics is the experiments on rats conducted by Michael Meaney of McGill University and subsequently by his former students who became researchers at Columbia University and elsewhere.[15] As mentioned in Chapter 1, in Meaney's lab, rat pups that were licked less by their mothers grew up to express symptoms of stress and neglected to lick their own pups. On the other hand, the pups that were licked frequently grew up to be good "lickers"; they licked their own young often, as did their offspring. Here was an instance in which attention and grooming (what humans might call "affection") seemed to have hereditable effects (as with the case of the Siberian foxes), and popularizers immediately seized upon the experiment and speculated that the same held true in humans. They postulated that neglected and abused children tended to become parents who neglected and abused their own children.

For the moment restricting ourselves to Meaney's rat experiments, we can ask, "What are the causes of such effects?" It is not obvious that they are genetic, since a form of learned behavior could be at work: pups learn from their mothers how to be good mothers. Meaney was well aware of this obvious hypothesis. He countered it with his finding that the adult offspring of good lickers had more glucocorticoid receptors in their brains than the offspring of poor lickers, making them more resistant to stress.

What could cause this? One of Meaney's colleagues, molecular biologist Moshe Szyf, was studying DNA methylation, particularly with regard to cancer. He and Meaney speculated that DNA methylation could act as an "on-off" switch for genes that affect the occurrence of cancer. Could something similar be at work with Meaney's rats? They studied Meaney's two groups of rats (groomed and ungroomed) and found different methylation patterns in their DNA. Their work pointed to an exciting hypothesis: a mother rat's licking could remove methyl groups from her pup's DNA. Therefore, the difference between the pups of grooming and nongrooming mothers might be an acquired characteristic and perhaps even reversible. However, in the absence of an explanation of the steps by which this happened, many biologists remained skeptical.

Both *Nature* and *Science,* the two leading science publications in the world, rejected Meaney and Szyf's research paper on these experiments.[16] The accepted view was that DNA methylation does not change after birth; this paper contradicted that view. Eventually, in 2004, *Nature Neuroscience* published the paper, and it caused a controversy.[17] But the case for epigenetic changes, not only for plants but now for animals, has grown stronger.[18]

Cross-fostering experiments with the rats strengthened (but did not prove) the view that epigenetics explained the different methylation patterns. If one takes the newborn pups of poor lickers and transfers them to the litters of good lickers, the pups of these "transferees" become good lickers as mothers. And the reverse is true: if one takes the pups of good lickers and transfers them at an early age to the litters of bad lickers, then the pups of those transferees are poor lickers as mothers.[19]

Could such effects be found in humans? In 2009 Meaney compared the brains of people who were abused as children and ultimately committed suicide to those of people who did not suffer such experiences and fates. The brains of those who had committed suicide showed methylation marks similar to those of the rats whose mothers were poor lickers.[20] Other researchers found supporting evidence in their studies of autism, asthma, and obesity, linking these conditions to pollution effects on children in Harlem and the South Bronx of New York City.[21]

This was heady stuff, and it led to great controversy. Meaney's publications became some of the most cited papers in *Nature Neuroscience*. Critics were quick to point out that Meaney's work was based on association, not "proof," and that neither Meaney nor anyone else knew the molecular processes by which environmental influences could cause methylation. Many links were missing, and therefore Meaney's views were often seen as provocative and challenging. Nonetheless, the scientific world was definitely paying attention to his work, his students' work, and increasingly, to other researchers following up on their leads.[22]

Another group of studies supporting the case for epigenetic inheritance in humans was based on the effects of famine. We obviously cannot submit humans to extreme experiences such as hunger just to find out whether there are any genetic effects. But sometimes, the cruelty of humans toward each other provides such experiments spontaneously, as was the case with the Dutch hunger winter of 1944–1945. In the fall of 1944, German forces were retreating before Allied assaults, but they still held many areas in the Netherlands, including Amsterdam. Dutch partisans fought the Germans behind the lines; in retaliation the Germans imposed a food embargo on the residents of these areas. That winter was by far the worst in recent Dutch history. By the end of February 1945, food rations dropped to about 580 calories a day in some areas (the normal consumption was about 2,300 calories daily for women and 2,900 for men). By May nearly 22,000 people had died of starvation. Almost all of the survivors, including pregnant women, suffered from severe malnutrition.[23] (Audrey Hepburn lived through the famine as a child and suffered from anemia and respiratory illnesses throughout her life.)

The Dutch are meticulous record keepers and have produced full statistics on deaths, births, disease, and disabilities among the population of this period; on the babies born during or shortly after the famine; and on the descendants of these children several generations later. The statistics on survivors and descendants are currently published under the rubric Dutch Famine Cohort Study.[24]

What these studies showed was surprising and pointed to epigenetic transgenerational inheritance. The grandchildren of mothers who were pregnant during the famine suffered from disabilities at abnormal rates, even though their own mothers were properly fed during their pregnancies. One author has called this "a grandmother effect."[25] These grandchildren suffered from many ailments at elevated rates: obesity, diabetes, coronary disease, breast cancer, lung and kidney problems, and depression.

Once again, critics pointed to weaknesses in these studies similar to those in Meaney's experiments with rats: they were all based on associations, not proof. It is well known to statisticians that association is far from being proof; on the other hand, association is a form of evidence, however weak, and the partisans of epigenetics continued to advance their arguments. Many associations pointing in the same direction began to mount up.

To the absolute amazement of Westerners, this rise in epigenetics and its link to the inheritance of acquired characteristics has led some Russian scientists to reconsider Lysenko and his views. Numerous recent publications have lauded Lysenko, insisting that his scientific claims were valid. In the next chapter, we will examine this rise of Lysenkoism and what it has meant for the field of Russian genetics.

8 | THE RECENT REBIRTH
OF LYSENKOISM IN RUSSIA

Now we can agree in part with Lamarck, although not very long ago this would be considered anti-scientific. And we see the countenance of Comrade Lysenko hanging over us.

—ALEKSEI BOIKO, *November 4, 2013*

IN THE 1990s and the first years of the current century, Russian scientists and science writers gradually became aware of the new field of epigenetics rapidly developing in the West. Inspired by the connection between epigenetics and the concept of the inheritance of acquired characteristics, a rebirth of Lysenkoism has occurred in Russia.[1]

Dozens of articles and books have appeared praising Lysenko and claiming that recent scientific developments have confirmed his views. One author called this phenomenon an attempted "exhumation of Lysenko."[2] If one goes to Yandex, the Russian equivalent of Google, and searches for "Lysenko" and "epigenetics" (in Russian), one will find a host of articles, books, and Internet blogs with titles such as:

"The Truth of Trofim Denisovich Lysenko is Confirmed
   by Modern Biology"[3]
"A Sensation: Academician Lysenko Turned Out to Be
   Right!"[4]

"Trofim—You are Right!"[5]

"Lysenko was closer to the truth than the Weismannites
   and Morganists"[6]

"Lysenko Was Right!"[7]

"In Memory of a Great Biologist—Lysenko"[8]

From the first moment, epigenetics was a controversial development in Russia, taking on a political character greater than in any
other country.[9] After all, Russia has a very special history in genetics. Lysenko had been discredited for a generation, but his name was
still inextricably linked to the concept of the inheritance of acquired
characteristics. Now, some epigeneticists were suggesting that such
inheritance was not only possible but increasingly documented. What
did this mean for Lysenko's reputation and standing?

The rise of epigenetics, though frightening to most established
Russian geneticists, was an opportunity for people nostalgic for the
old Soviet Union. A surprising number of remaining supporters of
Lysenko, many of them Stalinists, emerged. Most did not know much
about modern genetics, but a few who did pointed to new epigenetic research as the reason for the rehabilitation of Lamarck and
Lysenko.[10] And even one or two highly respected geneticists have
joined the Lysenko tide.

An article on this theme, which attracted great attention, appeared in the newspaper *Literaturnaia gazeta* (Literary Gazette), a
publication with a reputation in Soviet times for being admired and
read by Russian intellectuals. During the Soviet years, it even published, when permitted, articles criticizing Lysenko while he was in
power.[11] In 2002 the same newspaper published an article decrying
the appearance of so many articles about astrology, extrasensory phenomena, and sorcerers in the Russian press. The author, L. Korochkin, ridiculed these articles as "neo-Lysenkoism."[12]

Then, in 2009, when much of the world thought Lysenko was a tragic charlatan quickly being forgotten, the *Literaturnaia gazeta* published an article reviving him.[13] The author was Mikhail Anokhin, a pediatrician and specialist in pulmonology and physiology who after forty years of practice had branched out to become a publicist, short-story writer, and playwright. Calling himself a "doctor of biological sciences," Anokhin maintained that a revolution had recently occurred in genetics. He wrote that the form of genetics researched and propagated by Gregor Mendel, Thomas Hunt Morgan, and most biologists of the twentieth century was now "dead"; replacing it were the new sciences of molecular biology and epigenetics that he claimed were close to the Michurinist biology of Trofim Lysenko, which emphasized the role of the environment in inheritance.[14] Anokhin overlooked Lysenko's ignorance of molecular biology. He also ignored the fact that molecular biology and epigenetics grew out of classical genetics. Epigenetics supplemented classical genetics with new knowledge; it did not supplant it. The basic idea of epigenetics would not be possible without the work of Mendel, Morgan, and several generations of other researchers in the twentieth century describing "genes" located on specific spots on chromosomes.

In addition to misconstruing the history of genetics, Anokhin made other errors. He said, for example, that Lysenko's only big mistake was in believing that he could reconstruct human heredity on the basis on his theories. As we have seen, Lysenko always denied that his views extended to humans, restricting himself to plants and animals. Anokhin also said that Lysenko was overthrown when he criticized Khrushchev. In fact, Lysenko curried Khrushchev's favor and did not lose his position of authority until after Khrushchev had been deposed.

Anokhin claimed that Lysenko was far more important than other well-known Russian scientists such as Ivan Pavlov and Nikolai Vavilov. His article caused an enormous stir and elicited more than six hundred responses. However, most of the article's criticism appeared in other publications or in protest letters, not in the newspaper that originally published it.[15] The editors of the *Literaturnaia gazeta* showed little remorse about its publication. They contented themselves with publishing two responses, one favorable to Anokhin, another unfavorable, leaving the impression that the issue remained open.[16] And six years later, in 2015, the editors showed even more clearly where they stood by publishing yet another contribution by Anokhin, criticizing Vavilov and praising Lysenko.[17] Elsewhere, however, numerous critics raged against Anokhin. One author, Vladimir Gubarev, called Anokhin a "Devil from the Past" and "a liar." He further castigated the editors of *Literaturnaia gazeta* for "betraying all of us."[18]

Several leading geneticists responded. Academician Vladimir Gvozdev of the Institute of Molecular Genetics of the Russian Academy of Sciences said, "To analyze the illiterate and rambling article of M. Anokhin does not make sense" but then went on to analyze it. Anokhin had written that the cloning of the sheep Dolly and the genetic transposition work of Barbara McClintock, who received the Nobel Prize in 1983, validated Lysenko. Gvozdev responded by saying that both had "nothing to do with the anti-scientific views of Lysenko."[19] Another member of the Russian Academy of Sciences, Garri Abelev, was equally critical of Anokhin and affirmed that "we long ago dealt with Lysenko and any attempt to make him currently significant is out of the question."[20]

Despite his errors and misinterpretations, Anokhin struck a note that resonated in some social groups in Russian society of the early twenty-first century. With Putin's rise as the leader of Russia, ideas of Nationalism, nostalgia for Soviet times, and anti-Western sentiments were increasing. Finding in old Soviet figures like Lysenko a reason for claiming superiority over Western biologists accorded with these new trends.

It is difficult to imagine a more enthusiastic supporter of Lysenko than Iurii Mukhin.[21] According to Mukhin, Lysenko was an "eminent biologist of the twentieth century."[22] Mukhin stated that if the Michurinist biology supported by Lysenko had been allowed to develop fully, Russian biologists would not only have received many Nobel Prizes but might have found cures for cancer, AIDS, and diabetes.[23] Instead, according to Mukhin, the Soviet government in the 1960s gave in to the "stupid" views of the molecular biologists and persecuted Lysenko and his supporters, preventing them from continuing their work. The molecular biologists and classical geneticists who opposed Lysenko were, in Mukhin's opinion, "parasites" feeding on the necks of the Russian people and demanding high pay for their "useless" research. At one point he even called them "Jewish parasites."[24] The geneticists produced nothing of value, while Lysenko, according to Mukhin, brought immense benefit to Soviet agriculture. Mukhin described Soviet collective farms as "prosperous" and denied the famine of 1932–1933, in which millions died. To support these wild claims, Mukhin cited Soviet agricultural statistics that have long been discredited.[25]

Who is Iurii Ignat'evich Mukhin (1949–)? Educated as a metallurgist and a former head of a ferroalloy plant, he has no background in biology.[26] After the mid-1990s, he left technical work and became a political activist and a prominent Stalinist. He established and edited the Communist newspapers *Zavtra* (Tomorrow) and *Den'* (Day), followed by a newspaper titled *Duel'* (Duel) that was not only a font of Russian Nationalism but also of anti-Semitism. He lavished praise on Stalin and his murderous secret police chief, Lavrentii Beria. His views are so radical that in 2008 he was cited by a Russian court as an "extremist," but he continued his editorial work on the newspaper *K Bar'eru* (To the Barrier).[27]

Mukhin, by his own admission, knows little about genetics and molecular biology. In contrast to better-informed supporters of

Lysenko, he did not refer to epigenetics until 2014. Most of his criticism of "Weismannites-Morganists" are the same as those of Lysenko himself. According to Mukhin the followers of classical genetics in the Soviet Union did nothing of practical importance and wasted the money that the generous Soviet government gave them in the form of salaries and research facilities. He has maintained that Mendelism and the molecular genetics connected with it are "fruitless in essence."[28] He described Nikolai Vavilov, Lysenko's best-known opponent, as a foreign agent.[29] His major reference to events since Lysenko's reign has been his approval of Barbara McClintock's work.[30] He even claimed that Lysenko should have received the Nobel Prize given to McClintock because he anticipated her work by a "half-century."[31] However, he ignored the fact that McClintock considered herself a geneticist (she served as president of the American Genetics Society) and that she was thoroughly familiar with the modern genetics that Lysenko scorned. Mukhin, in contrast, has described genetics as "rubbish," "obscurantism," the "pride of stupid idiots," and a "prostitute" (the title of his book is *Genetics: A Prostitute*).[32]

In 2014 Mukhin suddenly caught up with epigenetics and celebrated what he saw as this new ally in his polemical struggle to restore Lysenko.[33] On March 27, 2014, the journal *Cell* published an article titled "Transgenerational Epigenetic Inheritance: Myths and Mechanisms," which mentioned Lysenko. *Cell,* originally founded at the Massachusetts Institute of Technology (MIT) and now published by Elsevier, is one of the most respected publications in biology, with a high impact factor.[34] The authors of the article, Edith Heard (Institut Curie, CNRS, France) and Robert Martienssen (Cold Spring Harbor Laboratory, United States) positively described Lysenko's "discovery of this cold-induced phenomenon in wheat and other cereals before the molecular basis of vernalization was known."[35]

American Barbara McClintock received the Nobel Prize in 1983 for her work on genetic transposition. (Science Source)

This was an erroneous description since, as we have seen, Lysenko's work on vernalization was not original at all. Furthermore, the genes responsible for the cold requirement of some plants became known while Lysenko was still in power, but he refused to recognize that work. But Heard and Martienssen did add that Lysenko "famously and unfortunately went on to propose that early flowering, induced by prolonged cold, could be inherited as an acquired trait. This led to disastrous attempts to rapidly breed high-yielding wheats that could be planted in the spring." Heard and Martienssen gave Lysenko only a little space in their long article and did not cite his works. Furthermore, Heard and Martienssen were generally skeptical of many claims of epigenetic inheritance, especially in animals, observing that "proof that transgenerational inheritance has an epigenetic basis is generally lacking in mammals."[36] Lysenko believed in the inheritance of acquired characteristics across the entire biological spectrum except humans and claimed that he had found it in many animals, especially dairy cows.

Mukhin ignored much of this and focused only on Heard and Martienssen's positive statement about Lysenko and vernalization. In an article on his website, "An Official Recognition of the Merit of Lysenko," Mukhin maintained that despite all the "lies poured on Lysenko's tomb," his scientific discoveries were beginning to be recognized. Mukhin wrote this commentary at the same time that Russian forces were taking over the Crimea and threatening Southern Ukraine. Mukhin, obviously hoping this effort would succeed, called for a monument to Lysenko to be erected in Odessa.[37]

Mukhin is not a serious participant in discussions of genetics, but the fact that his publications are winning some prominence in Russia is worrisome. The same article in *Cell* was picked up by the extreme right-wing weekly newspaper *Zavtra* in an article titled "Trofim—You Are Right!"[38] This paper is edited by Alexander Prokhanov, who in

1999 invited the American former Klansman David Duke to Russia to lecture. The author of the *Zavtra* article called the work of Heard and Martienssen an "affirmation of the correctness of Lysenko," which contradicts what the two Western authors actually wrote.

Vladimir Pyzhenkov has promoted Lysenko primarily by trying to denigrate his major opponent Nikolai Vavilov, a victim of repression.[39] Pyzhenkov has failed to recognize the fact that the validity of Lysenko's scientific views logically has nothing to do with the personal merits of Vavilov, or any of Lysenko's numerous critics. In Russia, however, the personal and class characteristics of the individuals involved in the struggles over Lysenko have often been emphasized more than the scientific elements of the arguments, a sad retreat to ad hominem attacks. It is particularly regretful that Pyzhenkov has followed this line of argument, since he was trained as a geneticist and has an academic background. But the defense of Stalinism trumps all other considerations in his 2009 book on Vavilov, whom Pyzhenkov described as the son of a wealthy merchant, a person inherently "anti-Soviet," and a traitor to the Soviet Union involved in espionage and the purposeful "wrecking" of Soviet agriculture.[40] In support of his views, Pyzhenkov cited the "confessions" that Vavilov made after imprisonment (and likely torture) to the secret police. Several generations of scholars, both Western and Russian, have discredited such forced confessions. Even the man in charge of Vavilov's interrogation in prison, Aleksandr Khvat, admitted many years later, after the demise of the Soviet Union, that "he didn't believe the spying charges against Vavilov."[41]

One of the most prominent Lysenko fans is a man writing under the name of Sigismund Mironin (a pseudonym; he has also written as A. Mirov). Mironin called Lysenko "the outstanding scientist of modern times."[42] Mironin earned a doctor's degree in biology and considers himself a professional biologist. In his 2008 book *Delo genetikov* (The Geneticists' Affair), his extensive bibliography includes 139 references, dozens in English. Look at his conclusions, given here verbatim:

1. Who was right: the Michurinists or the Morganists? The Michurinists and Lysenko were more correct.

2. Who was Lysenko? He was an outstanding natural scientist of the Soviet Union.

3. Is it true that Lysenko was a careerist who dishonestly stole his way into science and fooled Stalin? No.

4. Is it true that Lysenko was a scoundrel playing mean tricks? No. He was not a scoundrel, he did not denounce people, and did not play intrigues.

5. What did Lysenko discover that was new in biology and agricultural science? He made a whole series of outstanding discoveries.

6. Is it true that by means of organizing the August 1948 session of the Academy of Agricultural Sciences that Lysenko wanted to seize power and hold a monopolistic position in Soviet biology? No, he already had enough power, he simply wanted to defend himself against the Morganists.

7. Who first attacked, the Morganists or Lysenko? The Morganists attacked first. The 1948 session of the Academy of Agricultural Sciences was a defensive reaction of Stalin in answer to the attack of the Morganists.

8. Did Stalin support Lysenko? If, yes, why? Yes, Stalin supported Lysenko because he struggled against monopolism in science and against scientific clans.

9. What did Stalin want, organizing these open discussions? He wanted to make science more transparent, to free it from the clans.

10. Is it true that the monopoly of Lysenko had irreversible harmful effects on the development of Soviet biology? No, the influence of persecution was minimal.

11. If not, then why did Soviet biology fall behind? The basic cause of the retardation of Soviet biology and science as a whole was certain organizational errors, inadequate financing, and all-powerful princelings.[43]

This is a breathtaking list, written not only from a pro-Lysenko point of view but also a Stalinist one. Notice that Mironin said Stalin "wanted to make science more transparent." And throughout, the book describes Stalin as the man who made the Soviet Union strong, who created Soviet industry, who defeated Nazi Germany, who established scientific laboratories "no less outstanding" than those anywhere else in the world, and who should be praised for his policies toward science and education. According to Mironin, the purges and terror of the Stalin era have been grossly exaggerated by Western "cold warriors." Mironin has whitewashed not only Lysenko (who definitely attracted the attention of the secret police to biologists who opposed him) but also Stalin, one of the most murderous rulers of the twentieth century.

Defenses of Lysenko in Russia became much more serious in 2014 and 2015. In 2014 P. F. Kononkov, an old supporter of Lysenko, published a book titled *Two Worlds—Two Ideologies: On the Situation in the Biological and Agricultural Sciences in Russia in the Soviet and Post-Soviet periods.*[44] The book contained many of the standard arguments of Lysenko's supporters, but a new aspect alarmed the academic establishment in Russia: the publication was subsidized by a government organization, the Federal Agency on the Press and Mass Communication. The Russian historian of science Eduard Kolchinsky saw this subsidy as "the tolling of the bell" alerting Russian biologists to the specter of government support for Lysenko.[45]

Also in 2014 and 2015 two established biologists—Lev Zhivotovskii and A. I. Shatalkin—published books in which they described Lysenko as a significant scientist unappreciated in the West.

Zhivotovskii is a well-known scientist with a doctorate in biological science, a specialist in population genetics, and a researcher at the Institute of General Genetics of the Russian Academy of Science. He has published widely in international peer-reviewed journals on the topic of early human migration and has worked closely with foreign biologists, especially at Stanford University.

In his short 2014 book titled *The Unknown Lysenko,* Zhivotovskii raised Lysenko to the rank of "a great Soviet scientist" and based this claim on two arguments: First, that the early Lysenko "made great discoveries in plant physiology" and was "one of the founders of the biology of the development of plants."[46] And second, that the latest developments in biology, such as epigenetics, point to conclusions similar to Lysenko's and show his prescience as a scientist.

Although Zhivotovskii's credentials as a scientist garner our attention, neither assertion stands up to careful scrutiny. As shown earlier, Lysenko's work on the cold treatment and the staged development of plants was both repetitious of previous work (dating back centuries) and incredibly lacking in rigor. Repeating his experiments was almost impossible because of the laxness with which he described them and the small number of samples he used. Yet on these shaky bases, he erected grand hypotheses that have never been substantiated.

Zhivotovskii made much of the fact that a book on vernalization and photoperiodism, published in the United States in 1948, mentions Lysenko's work.[47] He saw that as international recognition of Lysenko's merit. However, a careful reading of the book shows the opposite: Lysenko's work comprised only a small part of a much larger world project studying vernalization and photoperiodism, and much of it was questionable. One author in the book noted that "no support . . . can be found for the view that spring cereals require higher temperature than winter cereals during vernalization, as Lysenko recommended."[48] Another remarked, "Neither does it appear probable that the method of vernalization will itself become widely used."[49]

Similarly, to conclude, as Zhivotovskii does, that epigenetics justifies Lysenko's theories is an enormous stretch. Why should Lysenko receive priority over dozens of previous biologists who defended the doctrine of the inheritance of acquired characteristics, including Russian colleagues who were very critical of Lysenko? The real test of Lysenko's views is not whether they drew on the inheritance of acquired characteristics but whether they have led to fruitful and continued research and application. That record is missing.

A. I. Shatalkin is an entomologist who has published widely in his field. He has long had an interest in Lamarck and the inheritance of acquired characteristics. In 2015 he published a book titled *Relational Conceptions of Inheritance and the Struggle about Them in the Twentieth Century*.[50] He began the book by saying that he would examine controversies over Lamarck and Lysenko in a very "objective" and calm way, trying to establish who was right and who was wrong. His conclusion is that both classical Mendelian genetics and the version of inheritance espoused by Lamarck and Lysenko are right, each in its own area. Classical geneticists are "correct," he says, in what they "affirm" but incorrect in what they "deny." They deny Lamarck's and Lysenko's valuable contributions to the science of heredity.

Shatalkin's claim to an objective and calm evaluation of rival theories of heredity is dramatically weakened in the later pages of the book, where he launched into an attack on Western genetics that reflects the new nationalism of the Putin era. He maintained that the West has launched a war against Russia in which Western scientists slander the "talented Lysenko" and describe themselves as preaching "the absolute truth." He insisted that Americans rail at Lysenko because they want to "cut off the wings of Russian science" and deprive Russia of the possibility of "internal development."[51]

Zhivotovskii's and Shatalkin's books are not the careful and well-founded examinations of Lysenko that they claim to be. Nonetheless, the entry into the rolls of contemporary defenders of Lysenko in Russia of such qualified scientists as Zhivotovskii and Shatalkin demonstrates that this phenomenon is serious and will not go away soon. Many established Russian geneticists feel concern; some now fear any research linked to Lysenko, especially in the new science of epigenetics.

9 | SURPRISING EFFECTS
OF THE NEW LYSENKOISM

Unfortunately, to my knowledge so far nobody looked for remote transgenerational (epigenetic) effects of this devastating hunger [the famine in the siege of Leningrad in World War II]. No one wanted to work on this because it might affirm Lamarckism and Lysenkoism.

—V. S. BARANOV

MANY OLD-LINE Lysenkoists and Stalinists trying to revive Lysenkoism in Russia are basing their efforts on the new science of epigenetics. They say it proves that Lysenko was correct (and, by implication, Stalin, since he supported Lysenko).

Established Russian geneticists, who know that Lysenko was a poor scientist, have been somewhat unwilling to explore transgenerational epigenetics because of their concern about the attempted rehabilitation of Lysenkoism. Given their experiences and history, they are a little frightened of epigenetics. As M. D. Golubovsky, a Russian biologist and historian of biology (now in the United States) wrote, "When a serious scholar found something that apparently conformed to Lysenko's views, he was afraid to make his discovery public, being scared of being ostracized by the academic community."[1] But epigenetics has also shaped a new and surprising attempt to connect Lysenko's views with Orthodox religion and has entered into distinctly political discussions about Russian attitudes toward government authority.

AVOIDING DISCUSSION OF TRANSGENERATIONAL
EPIGENETIC INHERITANCE

One of the best university textbooks on genetics in Russia today is Sergei Inge-Vechtomov's *Genetika s osnovami selektsii: Uchebnik dlia studentov vysshikh uchebnykh zavedenii* (Genetics and the Foundations of Selection: A Textbook for University Students), published in St. Petersburg in 2010. Its author is an internationally known and respected geneticist.[2] Born in 1939, he spent all of World War II in besieged Leningrad, which his father was killed defending. He graduated from Leningrad State University in 1961, during the last years of Lysenko's dominion of genetics, and he participated as a student and young scientist in the struggle against Lysenko. Today in his office at the same university, now named St. Petersburg State University, hang portraits of Russian geneticists who were imprisoned or executed because of their opposition to Lysenko. Inge-Vechtomov clearly remembers the struggles of his field and the importance of emancipation from Lysenko. His roots are in these authentic scientists' effort to rid their field of this oppressor.

After Lysenko was ousted from control in 1965, Inge-Vechtomov wanted to get fully in step with genetics worldwide, so he studied for two years in the United States, at both Yale University and the University of California, Berkeley. He returned to Russia determined to rebuild the field of genetics there. For many years he has been head of genetics at St. Petersburg State University, and he has published over 250 articles and books.

Inge-Vechtomov's university textbook discusses the latest research in many fields, including epigenetics. It describes how the methylation of DNA and the modification of histones (proteins that organize DNA strands and provide structural support) can affect gene expression. Nonetheless, a careful reader will notice that the discussion of epigenetics concentrates on these effects within one generation, not on transgenerational transmission of characteristics. The book does not deny such transmission but simply omits any large discussion of it. I suspect the reason for this is Inge-Vechtomov's aversion to anything that could be called "the inheritance of acquired characteristics." And knowing his history, I can fully understand.

FAMINE STUDIES

Although it is rare in modern history for a major city or area to suffer from prolonged famine, a few examples exist. One of the best-studied is the Dutch famine of 1944, which took place in the German-occupied Netherlands.[3] This research on the Dutch famine, which has been discussed more fully in Chapter 7, has, surprisingly, pointed to epigenetic changes during the famine that passed down through subsequent generations.

Other examples of the effects of food supply on future generations have been found in northern Sweden. The best-documented Swedish research on this topic is called the Överkalix Study. Överkalix is a region of several small villages in the far north of Sweden that in the early nineteenth century experienced terrible periodic food shortages. Interestingly, the same region sometimes enjoyed years of plenitude, when fish, grain, berries, and birds supplied an abundance of food. As a result, the isolated villagers of Överkalix starved during some years and gorged during other years. The effects of both experiences showed up in their grandchildren. Many parents of these grandchildren had moved from Överkalix to larger cities where nutrition was adequate all the time. Nonetheless, these grandchildren suffered from health defects even though they, and often their parents, had never experienced their grandparents' alternating famine and plenty.

According to the study, the grandsons of men who, just before puberty, were gluttons during seasons of plenty frequently died early. On the other hand, the granddaughters of women who starved when they were pregnant also frequently died early.

This research was so startling that the Swedish scientists who performed it, using the detailed health records of the people involved, were at first unable to publish their results. The leader of the research group was Lars Bygren of the Karolinska Institute, a noted medical university in Stockholm. Bygren himself was a descendant of an Överkalix village family. He reported, "It took many years, and we submitted to many journals. Scientists who reviewed and rejected the paper for publication did not quibble with the statistics. Rather they said 'it's impossible.'"[4] For such a thing to happen one would have to believe in "Lamarckian genetics."

It was only with the rise of epigenetics in the last fifteen years that Bygren's research was finally published.[5] His findings were suddenly seen as "possible" after scientists began to talk about how "epigenetic marks" caused by experiences such as famine might not get erased between generations and might be passed down, along with genes, for multiple generations.

If one is looking for examples of famines in recent history that might reveal information about epigenetics, the example of Leningrad in the period from 1941 to 1945 quickly comes to mind. As horrible as the periodic village famines in Sweden in the nineteenth century and the Dutch famine of 1944–1945 were, the Leningrad famine surpasses them immeasurably in terms of intensity and length. Only several thousand people lived in the Överkalix region of Sweden, and only several hundred died from famine. An estimated 22,000 died in the Dutch famine of 1944–1945. Incredibly, the famine deaths in Leningrad in 1941–1945 approached a million (the figures are disputed and range from 670,000 to 1.2 million). In Leningrad people

Women stripping meat from a dead horse during the siege of Leningrad in World War II. (Universal History Archive/Getty Images)

ate all the birds, rats, and pets in the city and then, in some cases, resorted to cannibalism. Many bodies taken to cemeteries were missing parts.

What were the lasting health effects of the Leningrad famine? Although Russian biologists made studies of the health of the immediate survivors, I can find no studies that discuss the transgenerational effects of the famine. This is striking, given the importance of other famines in the epigenetic literature. V. S. Baranov, an expert on the health effects of the Leningrad famine who as a small child lived through that event, told me on January 10, 2014, that "unfortunately to my knowledge so far nobody looked for remote transgenerational (epigenetic) effects of this devastating hunger. No one wanted to work on this because it might affirm Lamarckism and Lysenkoism."

Baranov himself has written articles studying the genetics of aging and longevity in survivors of the Leningrad siege but not the transgenerational effects. Another Russian medical biologist, Lidiia Khoroshinina, has done excellent work on the tragic effects in later life on children who suffered during the blockade but again, not on the transgenerational effects.[6] This omission, in my opinion, cannot be explained without referring to the reluctance of established geneticists in Russia to look at anything that might revive the concept of the inheritance of acquired characteristics. Just as famine leaves its scars on survivors, so also does the suppression of a science leave scars on scientific survivors.

EPIGENETICS AND ANIMAL DOMESTICATION

Another example of recent concerns about Lysenko and the inheritance of acquired characteristics comes from the research on the domestication of foxes near Novosibirsk, described in Chapter 1. When in 2007 several Western scholars suggested that epigenetics might help to explain the taming of the foxes, Russian biologists present at the conference called this thesis "dangerous" in its indication that Lysenko might have been partially correct.[7] Other biologists at Novosibirsk have been less afraid, and they have pointed out that Belyaev, the founder of the fox experiment, spoke of "dormant genes" and the effect of "stress" on genetic development, phenomena that might be explained today in terms of epigenetics.[8] At a banquet with the foreign advocates of epigenetics Eva Jablonka and Marion Lamb, a Russian even proposed a toast to Lamarck.[9] Other Russian geneticists present were not pleased, perhaps fearing that the next toast might be to Lysenko himself. These differing interpretations of the fox experiment reflect the traumatic experience of Lysenkoism in Russia.

LYSENKOISM AND RELIGION

A biology textbook very different from Inge-Vechtomov's that even more clearly reflects the stormy issues about biology raging in Russia today is S. Iu. Vert'yanov's 2012 *Obshchaia biologiia* (General Biology). This text, intended for the tenth and eleventh grades in Russian high schools, was published by the Trinity Lavra of St. Sergius, the most important Russian monastery and a spiritual center of the Russian Orthodox Church. Why is the Orthodox Church interested in biology? In recent years it has supported ever more strongly a Creationist view of the organic and inorganic world; this textbook is the church's effort to get such a view into the high school curriculum. Vert'yanov's text is thus devoted, first of all, to Creationism. In its foreword, Iu. P. Altukhov, professor emeritus of Moscow State University, has called the book "the first textbook of biology not restricted to materialistic viewpoints. We are returning to God in contrast to the last century of our life." And, indeed, the book depicts the world as God's creation and describes Darwin's views as doing "great damage not only to the development of science, but to humanity itself."[10] Throughout the book the Bible is quoted as a primary source.

Although this textbook is deeply religious, it is not ignorant of science. Similarly to some well-educated Creationists in the United States, the author has tried to both teach the facts of modern biology and encase them in a theological framework. The dominant themes in the book are Creationism, Nationalism, religion, anti-Darwinism, health (it strongly criticizes alcohol, cigarettes, and drugs), and ecology. On the latter theme, it says that because the world and all its creatures are the work of God, human destruction of that creation is immoral and sinful. It finds "consumer society" at fault for much of this destruction, and in its praise of Russian traditions, both religious and political, one catches glimpses of the continuing strength of the belief in the superiority of collectivism over individualism; it states that Darwin's views were in part based on the heartless competition of nineteenth-century British society.

Although Vert'yanov has not discussed epigenetics by name, he has referred to recent research showing that "some genes are more active than others depending on their interaction with other genes and the conditions of the environment." He has praised "Michurinist biology," the term Lysenko used for his ideas about inheritance. Since Vert'yanov is strongly critical of Darwin, any teaching that is implicitly critical of Darwin interests him. He sees the new emphasis on gene expression and the older ideas of Michurin as detrimental to Darwinism (conveniently forgetting that most epigeneticists still consider themselves Darwinists and that Michurin and Lysenko also praised Darwin). Vert'yanov has not gone all the way to the cementing of an alliance between Creationism and Lysenkoism, but he has strongly hinted at it.

Other writers supporting Lysenko have gone further in trying to cement a bond between Michurinist biology and Russian Orthodoxy. N. V. Ovchinnikov completely reversed a common interpretation of Lysenko.[11] Instead of seeing him as a Marxist, Ovchinnikov described him as a scientist whose views drew on deep Russian religious traditions. Ovchinnikov counterposed those traditions to the attitudes of Western pioneers of modern genetics such as Thomas Hunt Morgan, J. B. S. Haldane, and H. J. Muller, whom he described as "atheists" who believed genetics was a part of a blind, heartless, contingent Nature. Furthermore, Ovchinnikov castigated some of these pioneers, such as Muller and Haldane, for being both atheists and Marxists. Lysenko, according to Ovchinnikov, believed in a sacred natural order and man's dignified and exalted place in it. Ovchinnikov approvingly noted that C. H. Waddington, after talking to

Lysenko, concluded that "his philosophy has a strong taste of Orthodox religious theology."[12]

Ovchinnikov's interpretation is a stunning example of how Nationalist and Orthodox (Russian Orthodox) ideologies are reigning in Putin's Russia. Therefore, prominent Russians of the past, such as Lysenko, should be both national heroes and loyal to Orthodox religious thought. The fact that Muller and Haldane were sympathetic to atheism and Marxism or that Lysenko was Ukrainian, not Russian, are almost beside the point; what counted to Ovchinnikov was building a case for Russian Orthodoxy by belittling genetics.

Another Russian author who has emphasized Lysenko's link to orthodox theology and Russian traditions is P. F. Kononkov (mentioned in Chapter 8), a friend of Lysenko's who has steadfastly defended him.[13] According to Kononkov, Lysenko merely "cloaked" his views in Marxist, materialistic phrases because that was the requirement of the time. Kononkov has maintained that Lysenko was deeply religious, and his biological views reflected his religious roots. Kononkov agreed with Ovchinnikov that most of the Weissmannites and Morganists against whom Lysenko struggled were despicable atheists and Marxists. They believed that Lysenko was a defender of a religiously inspired natural order. He opposed artificial insemination and genetic manipulation, and he even spoke of "reproduction for love." Kononkov has gone on to launch a broadside attack on globalism, genetically modified organisms (GMOs), multinational companies, and the forces of rapacious capitalism, led by the United States, which, according to him, wished to wipe out all local differences. Deep in Kononkov's ideology is a romantic longing for a world unaltered by modernism. A plant specialist, he is also an herbalist and has promoted certain herbs (especially the plant amaranth) as a cure for various ailments.

EPIGENETICS AND HOMOSEXUALITY

Just as epigenetics has entered into current Russian disputes over religion and science, so also has it become implicated in discussions
about homosexuality. Compared to most Western nations, Russia still
displays a marked intolerance of homosexuality. The government
enacted a law that prohibits "propaganda of non-traditional sexual
practices" and has also barred demonstrations by gays and lesbians.
Epigenetics has recently been caught up in these issues.

In December 2012 a trio of researchers, two from the United
States and one from Sweden, published a technical article in the *Quarterly Review of Biology* titled "Homosexuality as a Consequence of
Epigenetically Canalized Sexual Development," in which they pre-
dicted that "homosexuality is produced by transgenerational epigenetic inheritance." In their opinion, "homosexuality occurs when
one or more stronger than average epi-marks . . . carryover across
generations into an opposite-sex descendant."[14]

This article caused a considerable stir in many countries. It was
picked up in the United States by CNN, *Time, Science,* Fox News, and
many newspapers. It was also featured in a number of prominent
Russian publications. The Russian media were particularly attracted
to the issue by the fact that one of the authors of the article was a
Russian-American, Sergey Gavrilets, who had received his higher
education at Moscow University.

The distinctly different interpretations of the article given in the
press in the United States and Russia are striking. In the United States,
the typical headlines stated: "Homosexuality Can Be Passed Down"
(CNN) or "Homosexuality is Epigenetic, Says Study" *(Science).* In
Russia, typical headlines were: "Mistakes in Cell Memory May be
Cause of Homosexuality" *(RIA Novosti)* or "Gays and Lesbians are
Created by Mistakes" *(Komsomolskaia pravda),* emphasizing the word

"mistake."[15] In other words, the Russian press gave the impression that even if "nontraditional" sexual choice is influenced by genetics, it is only through genetic "mistakes" and therefore, not normal.

If one examines the original article, one sees that the Russian interpretation was misleading. Not once in the *Quarterly Review of Biology* article did the trio of authors use the word "mistake." Instead, they described what they saw as epigenetic changes contributing to sexual preference occurring "by happenstance and with moderate to low probability."[16] To them, it was a normal process.

EPIGENETICS AND POLITICS

Yet another developing controversy in Russia about epigenetics concerns politics, in particular, the legacy of Stalinism in political behavior. A few of the scientists in Russia who are outraged by the recent rise of Lysenkoism have concluded that their best defense, relying on recent research, is to take the weapon of epigenetics away from the Stalinist Lysenkoists. They are saying to supporters of Lysenko, in effect, "Let us agree that acquired characteristics can in some instances be inherited, and let us also agree that epigenetics is very important for the understanding of recent history. What would that mean for you Stalinists?" And the answers they give are designed to embarrass, if not discredit, the Stalinists.

Such a person is Vladimir Kozlov, director of the Institute of Clinical Immunology of the Russian Academy of Medical Sciences. Kozlov has cited the research of Brian Dias and Kerry Ressler at Emory University, maintaining that rats can epigenetically inherit the fear of certain smells if those same smells in an earlier generation were associated with negative experiences, such as electrical shocks.[17] Using this example, Kozlov believes that political repressions during the Stalinist years lamed the Russian people epigenetically.[18] During

the 1930s, 1940s, and early 1950s several million Russian citizens
were imprisoned, many thousands were executed, and the entire
population was subjected to fear. According to Kozlov, epigenetics
shows that it takes three to five generations for organisms to escape
the effects of such widespread fear, and therefore the present Russian
population still suffers from Stalinist repressions. In his opinion,
"fear for themselves and their families" explains Russians' political
passivity and willingness to tolerate authoritarian rulers. Another
Russian author wrote that epigenetics explains the "suicidal submis-
siveness" of the Russian people.[19]

What these emerging disputes in Russia today—over Lamarckist
explanations of animal domestication, over the legitimacy of certain
kinds of biological research, over religion and science, over the effects of political repression, and over homosexuality—show is that
epigenetics is being used as a football by conflicting ideological fac-
tions. Proving assertions about a relationship between epigenetics and
political fear in Russian society today, or that epigenetics has a
direct relationship to homosexuality, is elusive. Even proving the
role of epigenetics in the domestication of animals is, at the present
stage of science, beyond our ability. Speculation usually takes over.
These discussions say a lot more about Russian politics than they do
about science.

Russia, with its distinct and painful history in genetics, displays
political and biological passions about epigenetics more clearly than
any other nation. The rather dramatic acceptance of the concept of
the inheritance of acquired characteristics, discredited during much
of the last century, complicates matters. But as Chapter 10 will reveal, scientists have been able to unlink epigenetics from the science
of Trofim Lysenko.

10 | ANTI–LYSENKO RUSSIAN SUPPORTERS OF THE INHERITANCE OF ACQUIRED CHARACTERISTICS

> A great majority of Mendelists truly denied the possibility of the inheritance of acquired characteristics, considering that doctrine to be out-dated Lamarckism, and in this way, in my opinion, they were mistaken. . . . Lysenko and his allies asserted that the chromosome theory and Mendelism were so untrue and fruitless that they should never be used. And this position was simply untrue.
>
> —A. A. LIUBISHCHEV

PERHAPS THERE IS A WAY to accept the concept of the inheritance of acquired characteristics without linking it to Lysenko's work. Ironically, scientists have been more successful at doing this in Russia than in the West.

In Chapter 1, I related Dmitri Belyaev's reaction when I asked him over thirty years ago if he knew that some of his laboratory assistants at his fox farm were Lysenkoists because they believed in the inheritance of acquired characteristics. Belyaev laughed and replied that he knew about their views, contrary to his own. He insisted that his assistants were not Lysenkoists but simply supporters of the idea of the inheritance of acquired characteristics. In order to answer the larger question of whether Lysenko should be given new credit in

light of the rise of epigenetics, it is helpful to return to Belyaev's reaction. He knew that the doctrine of the inheritance of acquired
characteristics and Lysenkoism were not the same, and he knew that
it is critical to make this distinction.

In recent conversations with other academic geneticists in Russia,
I have found that many of them also emphasize this difference. In
fact, they speak of it much more than academic geneticists in the West
usually do, particularly in the United States. A root cause is that in
Russia, the belief in the inheritance of acquired characteristics was
supported by many biologists other than Lysenko and never died out
as it did in the United States. Further, some of Lysenko's most dedicated critics, people who suffered because they opposed him, were
proponents of the inheritance of acquired characteristics. Russian authors never uniquely tied the concept to Lysenko, as many did outside the Soviet Union. This fact has been almost totally overlooked
in the West, but it is now, after the advent of epigenetics, crucial to
understanding the significance of Lysenko.

We should note that L. I. Blacher, a Russian author who in 1971
wrote a well-researched book titled *The Problem of the Inheritance of
Acquired Characters,* gave scant attention to Lysenko. There were plenty
of other Russian biologists much more scientifically qualified than
Lysenko, who had been attracted to the concept. They included S. G.
Levit, E. S. Smirnov, N. D. Leonov, B. S. Kuzin, and, later, A. A.
Liubishchev. In the chapter titled "Discussions in the Soviet Union
Concerning the Inheritance of Acquired Characters: 1920s," Blacher
did not mention Lysenko once, even though Lysenko was active and
publishing from 1923 onward.[1]

Prominent entomologist A. A. Liubishchev (1890–1972), a supporter of the inheritance of acquired characteristics, was one of Lysenko's bravest opponents, writing from 1953 to 1964 more than one
thousand manuscripts lambasting Lysenko. None were published at

the time—and could not be published—but they circulated widely as *samizdat* (underground self-published) documents.[2] Liubishchev also wrote letters to the head of the Soviet government, Nikita Khrushchev, and other Soviet political leaders calling for the elimination of Lysenko's domination of genetics.[3] As one of Liubishchev's friends noted, his writings were "widely known to a small circle of influential readers."[4]

Although many Western observers assumed that the Lysenko controversy centered on the inheritance of acquired characteristics, to Liubishchev that was not the issue at all. He accepted that doctrine.[5] He even said that the classical geneticists in the Soviet Union who were vanquished by Lysenko erred in denying the inheritance of acquired characteristics and in trumpeting chromosomes and DNA as the only carriers of heredity.[6] Liubishchev was convinced that heredity was more complicated than that.

If Liubishchev agreed with Lysenko about the inheritance of acquired characteristics, why did he risk his life by waging an all-out battle against him? Liubishchev believed that the essential "evil" (his word) in Lysenko's approach to science was "intolerance, dogmatism, and ignorance." Lysenko was a "falsifier and a deceiver."[7] He brooked no rivals, worked with government officials to eliminate his critics, demanded that biology textbooks and university lecture courses be written to his prescription, and failed to follow elementary standards of statistics and reproducibility in his research. He achieved his goals because of state power. Lysenko practiced, said Liubishchev, a "closed science" reminiscent of the worst scholasticism of the Middle Ages, and he treated his opponents like it was the Spanish Inquisition.[8] Liubishchev saw Lysenko as an enormous step backward for Russian science, and he used all of his oratorical talents to try to convince Soviet leaders to allow a plurality of voices in Soviet biology. Liubishchev also pleaded with his fellow geneticists to make the main target of their attack on Lysenko his careless, sloppy methods of research, not his particular conclusions. Liubishchev believed that because of Lysenko's dishonesty and lack of standards his results should not be considered legitimate.

One of Liubishchev's most interesting articles was written in 1965, *after* the fall of Lysenko.[9] At a time when geneticists all over Russia were celebrating victory over a person they saw as a vicious dictator in their field, Liubishchev took a more sober position. Although he also rejoiced at the overthrow of Lysenko, he warned of the dangers of placing "blind faith" in classical genetics and excluding all considerations of the inheritance of acquired characteristics. He criticized the "intolerance and arrogance" of some defenders of classical genetics who were now triumphing over Lysenko. Anticipating criticism of his view, Liubishchev asked himself whether the "tolerance" he favored meant that he still saw a place for Lysenkoism alongside classical genetics. No, he replied to his own question, because Lysenko did not meet the standards of research methodology (statistics, reproducibility, transparency, and reporting all research results, both favorable and unfavorable). But proponents of the inheritance of acquired characteristics who did meet these standards, Liubishchev said, should be included in the scientific community respected by all.[10]

Liubishchev considered himself to be one of those proponents. Thus, he made a major effort to separate the inheritance of acquired characteristics from Lysenkoism and reveal Lysenko's harmful dogma. For years Liubishchev was unsuccessful, but some people think he played an important role in weakening government support for Lysenko. So as you can see, it is possible to accept the inheritance of acquired characteristics *without* bringing Lysenko back.

CONCLUSION

WITH THE REALIZATION that the inheritance of acquired characteristics might happen after all, was Lysenko right? No, he was not. Some people may think so because they mistakenly link Lysenko uniquely to the doctrine of acquired characteristics, a belief that has been around for several thousand years. Lysenko was a very poor scientist, and the inheritance of acquired characteristics was actually a small part of what he claimed.

The fathers and mothers of epigenetics did not use Lysenko's results but developed their views on the basis of molecular biology.[1] The concepts of gene expression and the methylation of DNA would have been impossible without intimate knowledge of molecular biology. Rather than denying the importance of the genome, epigeneticists state that classical genetic information makes the environmental effects of gene expression possible. The environment regulates transcription factors that bind to regulatory elements on the DNA and activate or repress gene expression—but the nucleotide sequence determines the ability of the transcription factors to bind in the first place. Epigeneticists emphasize the importance of the genome but describe it as more "permissive" than earlier thought.[2]

Lysenko could not possibly agree with the description of epigenetics given in the previous paragraph. He disregarded the action of genes, and in 1974, two years before his death, made the following amazing statement:

> I declare that we have never used and are not going to use any ideas and methods of molecular biology. I would like to advise all biologists, plant and animal breeders and students

in the Soviet Union against adopting these methods, as they only hinder our understanding of the essential, that is advancement of theoretical biology.[3]

This was a man who would throw biology back decades, if not centuries. Warren Weaver coined the term "molecular biology" in 1938; in 1974, when Lysenko made the above statement, it was blossoming all over the world. To give Lysenko credit for what the pioneers of epigenetics did through enormous labor, based on the latest developments in molecular biology, would be both inaccurate and unjust.

Does this mean that Lysenko was totally worthless as a practical plant breeder, especially in his early years? No. Lysenko had talents in the field, and some of his concepts, such as graft hybridization, still deserve consideration (although it was not new; Charles Darwin, Ivan Michurin, Luther Burbank, and others promoted the idea). If Lysenko had lived in a normal democratic country, he would be remembered, if at all, as a talented farmer working away in his fields, employing idiosyncratic methods but never garnering much support. None of his methods are employed in Russia today. But in the Soviet Union in the 1930s, a country suffering from famine (caused in large part by the disastrous collectivization effort), the need for quick agricultural remedies was acute, and Lysenko offered them. No aspiring promoter of a peculiar scientific system ever fell into a more personally fortunate (and historically tragic) situation. The relationship between Lysenko and his political environment was one of reciprocal corruption. As C. D. Darlington commented:

His modest proposals were received with such willing faith that he found himself carried along on the crest of a wave of disciplined enthusiasm, a wave of such magnitude as only totalitarian machinery can propagate. The whole world was overwhelmed by its success. Even Lysenko must have been surprised at an achievement which gave him an eminence shared only by the Dnieper Dam.[4]

Of course, the statistics on crop yields used to support Lysenko were false and manipulated, as were most Soviet statistics of the time. The undereducated and narrow-minded Lysenko was carried away and transformed by the adulation he received. Infatuated with his success, he fought bitterly and underhandedly against anyone who opposed him. He learned that his most powerful approach was to advertise himself as a simple peasant, not a member of the Communist Party, while denouncing his critics as remnants of the tsarist bourgeoisie who were opposed to everything Soviet. At a time when arrests of "bourgeois specialists" by the hundreds were occurring all over the Soviet Union, it was fairly easy for Lysenko to get rid of his opponents.

But let us, momentarily, overlook Lysenko's moral and political failings and ask a simple question: Was Lysenko right in at least some of his scientific views? My answer is the following: where he was right, he was not original; where he was original, he was not right. He was right in his belief in the inheritance of acquired characteristics, but so were many of his predecessors and contemporaries. He was original in his claims to change one species into another, but his claims have not been replicated, and we must conclude that he was wrong. Of course, the change of one species into another must have occurred in the biological world, or we could not have evolution. And in recent years, by means of genome transplantation, it has evidently been done in bacteria.[5] But to attribute this achievement to Lysenko would be as mistaken, in my opinion, as to attribute the success of epigenetics to him. In both cases he was simply outside the science that would have permitted him to make these advances.

Lysenko was actually a very poor representative of the concept of the inheritance of acquired characteristics. Other scientists, both in Russia and elsewhere, did a far better job. Lysenko's experiments were careless and usually unverifiable. The inheritance of acquired characteristics was better defended, even during Lysenko's reign, by scientists such as the American Tracy Sonneborn, who demonstrated such inheritance in the protozoan group *Paramecium*.[6]

What will result from the rise of neo-Lysenkoism in Russia? Is there a genuine threat to research biologists in that country? Could Lysenkoism take over Russia again? The possibility of such a cataclysm happening is very small. Russia under President Vladimir Putin is distressingly authoritarian, but it is not the same as Stalin's Soviet Union in terms of the degree of control over society. In the Soviet Union of Lysenko's time, the Communist Party and the secret police controlled every university, every research institute, every academic journal, and every newspaper.[7] No one could travel abroad without the permission of these authorities. Imposing a particular view in biology was a question of state policy. The scientists in Putin's Russia are in constant contact with their Western colleagues, frequently traveling to meet them in professional gatherings and working together in cooperative research projects. The Internet and e-mail have bound the world's scientists together. Support for Lysenko among Russian scientists engaged in such international cooperation is very small.

Furthermore, Putin's government wants Russia to become a leading high-technology country. Biomedical research is one of the areas of research featured in Skolkovo, the current attempt to duplicate Silicon Valley. Without full knowledge and excellence in molecular biology, which is based on principles very different from

Lysenko's biology, eminence in biomedicine will not be possible. Talented researchers in Russian biology understand this very well. Their administrators understand it also. And even Putin and his associates understand it. Thus, little danger exists that Lysenkoism will again take over academic genetics in Russia. Instead, the threat is that Lysenko's supporters will influence public perceptions and perhaps even secondary education. As we have seen, Nationalists have produced a new biology textbook for tenth and eleventh graders that represents these views, and they are pushing for its adoption in local schools.

This neo-Lysenkoism is perpetrating damage of a different sort: it is warping our understanding of the past. Not only Russia's past, but ours too. Unfortunately, many people, both in Russia and in the West, are willing to accept an interpretation of Lysenko that goes something like this: "Lysenko was a nasty man, but we should grant that he was right about the inheritance of acquired characteristics and therefore give him more credit than we have in the past." Examples of such interpretations are appearing in Western journals.[8] Yet the new biological understandings that we now possess, including epigenetics, do not originate from anything that Lysenko did; they arose out of the classical genetics that he spurned. Lysenko deserves far less credit for defending the doctrine of the inheritance of acquired characteristics than dozens of other researchers who opposed him while supporting this doctrine. But instead of knowing the names of these scientists, we know Lysenko's name because of his tyranny.

Several times in this book, I have written of the victory of "usage" over "accuracy." We tend to reduce the scientists of the past to one concept or thought, ignoring the rest of their work. Thus, when we mention Lamarck, we think "inheritance of acquired characteristics," even though this concept was actually a small part of his opus and was shared by most of his colleagues. When we mention

Weismann, we think "germ plasm theory," although he actually equivocated on this theory. And now we face the question of Lysenko. When his name comes up today, do we think "inheritance of acquired characteristics" or "incompetent scientist who, with the help of state repression, politically imposed his views on others"? It is an important question. The latter interpretation is, in my opinion, the valid one. No, Lysenko was not right after all.

NOTES

INTRODUCTION

1. Michael Gordin, "How Lysenkoism Became Pseudoscience: Dobzhansky to Velikovsky," *Journal of the History of Biology* 45, no. 3 (2012): 443–468.

2. For example, Nessa Carey, *The Epigenetics Revolution: How Modern Biology Is Rewriting Our Understanding of Genetics, Disease, and Inheritance* (London: ICON, 2012).

3. For example, Edith Heard and Robert A. I. Martienssen, "Transgenerational Epigenetic Inheritance: Myths and Mechanisms," *Cell* 157 (March 27, 2014): 95–109. Note their comment on p. 95 that "proof that transgenerational inheritance has an epigenetic basis is generally lacking in mammals" and their further conclusion on p. 106 that "environmentally-induced epigenetic changes are rarely transgenerationally inherited, let alone adaptive, even in plants. Thus, although much attention has been drawn to the potential implications of transgenerational inheritance for human health, so far there is little support."

4. David Joravsky, *The Lysenko Affair* (Cambridge, MA: Harvard University Press, 1970); Loren R. Graham, "Genetics," in *Science and Philosophy in the Soviet Union* (New York: Knopf, 1972), 195–256; Zhores Medvedev, *The Rise and Fall of T. D. Lysenko* (New York: Columbia University Press, 1965); Valery N. Soyfer, *Lysenko and the Tragedy of Soviet Science* (New Brunswick, NJ: Rutgers University Press, 1994); Nikolai Krementsov, *Stalinist Science* (Princeton, NJ: Princeton University Press, 1997); Dominique Lecourt, *Proletarian Science? The Case of Lysenko* (Atlantic Highlands, NJ: Humanities Press, 1977); Nils Roll-Hansen, *The Lysenko Effect: The Politics of Science* (Amherst, NY: Humanity Books, 2005); William deJong-Lambert, *The Cold War Politics of*

Genetic Research: An Introduction to the Lysenko Affair (Dordrecht: Springer, 2012). An early insubstantial and politicized account is James Fyfe, *Lysenko Is Right* (London: Lawrence and Wishart, 1950). See also Nikolai Krementsov, "Marxism, Darwinism and Genetics in the Soviet Union," in *Biology and Ideology: From Descartes to Dawkins,* ed. Denis R. Alexander and Ronald L. Numbers (Chicago: University of Chicago Press, 2010), 215–246; and Nikolai Krementsov, *International Science between the World Wars: The Case of Genetics* (London: Routledge, 2005).

1. THE FRIENDLY SIBERIAN FOXES

Parts of the discussion in this chapter rely on my book *Moscow Stories* (Bloomington: Indiana University Press, 2006) and on my role as rapporteur for a NOVA science program in 1986 that featured Belyaev's farm.

Epigraph: V. Bartel', "90-letiiu akademika D. K. Beliaeva posviashchaetsia," *Nauka v Sibiri* 34 (September 6, 2007): 8–9.

1. Paul R. Josephson, *New Atlantis Revisited: Akademgorodok, the Siberian City of Science* (Princeton, NJ: Princeton University Press, 1997).

2. One of them was Nikolai Vorontsov, who later, as minister of the environment, became the only non-Communist member of the Council of Ministers in the history of the Soviet Union. He later told me jokingly that his ambition was to become the first nonmember of the Central Committee of the Communist Party.

3. Alexei V. Kouprianov, "The 'Soviet Creative Darwinism' (1930s–1950s): From the Selective Reading of Darwin's Works to the Transmutation of Species," *Studies in the History of Biology* 3, no. 2 (2011): 8–31.

4. See T. D. Lysenko, ed., *I. V. Michurin: Sochineniia v chetyrek tomakh,* vols. 1–4 (Moscow: Gosizdat, 1948).

5. Richard C. Francis, *Domesticated: Evolution in a Man-Made World* (New York: W. W. Norton, 2015).

6. The fox farm experiment was featured in *National Geographic:* Evan Ratliff, "Taming the Wild," *National Geographic* (March 2011), http://ngm .nationalgeog-wild-animals/ratliff-text.

7. I. C. G. Weaver et al., "Epigenetic Programming by Maternal Behavior," *Nature Neuroscience* 7, no. 8 (August 2004): 847–854.

8. F. A. Champagne et al., "Maternal Care Associated with Methylation of the Estrogen Receptor–Alpha1b Promoter and Estrogen Receptor–Alpha Expression in the Medial Preoptic Area of Female Offspring," *Endocrinology* 147 (June 2006): 2909–2915.

9. I am grateful to Everett Mendelsohn, who put me in contact with Eva Jablonka, who in turn told me about her visit to Novosibirsk. She presented there, with Marion J. Lamb, the paper "Epigenetic Inheritance as a Mediator of Genetic Changes during Evolution." See also Eva Jablonka and Marion J. Lamb, *Epigenetic Inheritance and Evolution: The Lamarckian Dimension* (Oxford: Oxford University Press, 1995).

10. Bartel', "90-letiiu Akademika D. K. Beliaeva posviashchaetsia," 8–9.

11. Julia Lindberg et al., "Selection for Tameness Has Changed Brain Gene Expression in Silver Foxes," *Current Biology* 15, no. 22 (2005): R915–916.

2. THE INHERITANCE OF ACQUIRED CHARACTERISTICS

Epigraph: Conway Zirkle, "The Early History of the Idea of the Inheritance of Acquired Characters and of Pangenesis," *Transactions of the American Philosophical Society,* n.s., 35, pt. 2 (1946): 91.

1. See Jessica Wang, *American Scientists in an Age of Anxiety: Scientists, Anticommunism, and the Cold War* (Chapel Hill: University of North Carolina Press, 1999).

2. M. D. Golubovsky, *Vek genetiki: Evoliutsiia idei i poniatii* (St. Petersburg, Russia: Borei Art, 2000), 167.

3. G. L. Stebbins, *Darwin to DNA, Molecules to Humanity* (San Francisco: Freeman, 1982), 76.

4. *New World Encyclopedia,* http://www.newworldencyclopedia.com.

5. Scott Kennedy, "Multigenerational Epigenetic Inheritance" (lecture, Harvard Medical School, Boston, MA, December 4, 2013). I am grateful to Dr. Felton Earls for inviting me to the lecture.

6. Zirkle, "The Early History of the Idea of the Inheritance of Acquired Characteristics and of Pangenesis," 91.

7. See Richard W. Burkhardt Jr., *The Spirit of System: Lamarck and Evolutionary Biology* (Cambridge, MA: Harvard University Press, 1977).

8. See Stephen J. Gould, *The Structure of Evolutionary Theory* (Cambridge, MA: Belknap Press of Harvard University Press, 2002), 170–174.

9. See Frederick B. Churchill, *August Weismann: Development, Heredity, and Evolution* (Cambridge, MA: Harvard University Press, 2015).

10. R. G. Winther, "August Weismann on Germ-Plasm Variation," *Journal of the History of Biology* 34 (2001): 530, 550.

11. See P. Chalmers Mitchell, "The Spencer-Weismann Controversy," *Nature* 49 (February 15, 1894): 373–374.

12. Karl Frantsevich Rul'e, "Obshchaia zoologiia," in *Izbrannye biologicheskie proizvedeniia* (Moscow: Izdatel'stvo AN SSSR, 1954), 127–128.

13. L. I. Blacher, *The Problem of the Inheritance of Acquired Characters* (New Delhi: Amerind, 1983), 102–109.

14. See Daniel Todes, *Ivan Pavlov: A Russian Life in Science* (New York: Oxford University Press, 2014), 450–463.

15. Ibid., 451.

16. I follow here the careful translation distinction between "conditional" and "conditioned" made by Daniel Todes.

17. Todes, *Ivan Pavlov,* 457.

18. T. H. Morgan, "Are Acquired Characters Inherited?," *Yale Review,* July 1924, 712–729, cited in Todes, *Ivan Pavlov,* 460.

3. PAUL KAMMERER, ENFANT TERRIBLE OF BIOLOGY

Epigraph: Quoted in Arthur Koestler, *The Case of the Midwife Toad: A Scientific Mystery Revisited* (New York: Random House, 1971), 63.

1. Ibid., 91.

2. D. K. Noble, "Kammerer's *Alytes*," *Nature* 18 (August 7, 1923): 208–210.

3. Koestler, *Case of the Midwife Toad.*

4. Klaus Taschwer, science editor of *Der Standard,* is working on this theme. See his "The Toad Kisser and the Bear's Lair: The Case of Paul Kammerer's Midwife Toad Revisited," Max-Planck-Institut für Wissenschaftsgeschichte, on the institute website, 4.2011–6.2012, https://www.mpiwg-berlin.

5. Elizabeth Pennisi, "The Case of the Midwife Toad: Fraud or Epigenetics?" *Science* 325 (September 4, 2009): 1194–1195.

6. Ibid., 1195.

7. Günter P. Wagner, "Paul Kammerer's Midwife Toads: About the Reliability of Experiments and Our Ability to Make Sense of Them," *Journal of Experimental Zoology* 312B (2009): 665–666; see also Gerald Weissmann, "The Midwife Toad and Alma Mahler: Epigenetics or a Matter of Deception?," *FASEB Journal* 24 (August 2010): 2591–2595. Weissmann considers Kammerer a fraud.

8. Paul Kammerer, *The Inheritance of Acquired Characteristics* (New York: Boni and Liveright, 1924), 31.

9. Ibid., 263.

10. Of course, some scholars have studied left-wing Darwinism, such as Piers J. Hale in his *Political Descent: Malthus, Mutualism, and the Politics of Evolution in Victorian England* (Chicago: University of Chicago Press, 2014). See also Diane Paul, "Eugenics and the Left," *Journal of the History of Ideas,* no. 4 (1984): 567–585. Even more appropriate to this book is M. Meloni, *Political Biology: Social Implications of Human Heredity from Eugenics to Epigenetics* (London: Palgrave, 2016).

11. "Poor Are Able to Purchase Soft Pillows," in Kammerer, *Inheritance of Acquired Characteristics,* 269–270; "White 'Caucasians' Who Live Long Enough," in ibid., 279; "Certain Physical and Psychical Traits," in ibid., 274.

12. Daniel Todes, e-mail message to author, May 7, 2014. Some people have speculated that Ivan Pavlov may have played a role in inviting Kammerer to come to Russia. Pavlov did display interest in the inheritance of acquired characteristics, but there is no evidence that Pavlov actively supported Kammerer. I am indebted to Pavlov's biographer Daniel Todes for this information. For a description of Kammerer's Russian tour, see Margarete Vöhringer, "Behavioural Research, the Museum Darwinianum and Evolu-

tionism in Early Soviet Russia," *History and Philosophy of the Life Sciences* 31 (2009): 279–294; and P. Kammerer, "Das Darwinmuseum zu Moskau," *Monistische Monatshefte* 11 (1926): 377–382.

13. See mosintour.ru/paul_kammerer.

14. Anatoli Lunacharskii, *Lunacharskii o kino: Stat'i, vyskazyvaniia, stsenarii, dokumenty* (Moscow: Izdatel'stvo iskusstvo, 1965), 151–155.

15. Ibid., 151.

16. Ibid., 152.

17. Anatoli Lunacharskii, *Lunacharskii o kino* (Moscow: Izdatel'stvo iskusstvo, 1965), 4, http://www.plantphysiol.org/content/138/4/2364.full.

18. Ibid., 5.

19. Koestler, *Case of the Midwife Toad*, 94.

20. See Denise J. Youngblood, "Entertainment or Enlightenment? Popular Cinema in Soviet Society, 1921–1931," in *New Directions in Soviet History*, ed. Stephen White (Cambridge: Cambridge University Press, 2002), 41–50.

21. Lunacharskii, *Lunacharskii o kino*, 153.

22. *Jump Cut: A Review of Contemporary Media* 26 (December 1981): 39–41.

23. F. Lenz, "Der Fall Kammerer und seine Umfilmung durch Lunatscharsky," *Archiv für Rassen-und Gesellschafts-Biologie* 21 (1929): 316.

24. See Lenz's review of Goldschmidt's *Einführung in die Vererbungswissenschaft*, in *Archiv für Rassen-und Gesellschafts-Biologie* 21 (1929): 99–102.

25. Koestler, *Case of the Midwife Toad*, 144.

26. "Kammerer," *Meditsinskaia entsiklopediia*, in http://medencped.ru/kammerer/.

4. THE GREAT DEBATE ABOUT HUMAN HEREDITY IN 1920S RUSSIA

Parts of this chapter rely on my "Eugenics: Weimar Germany and Soviet Russia," in my book *Between Science and Values* (New York: Columbia University Press, 1981).

Epigraph: Paul Kammerer, *The Inheritance of Acquired Characteristics* (New York: Boni and Liveright, 1924), 364.

1. N. K. Kol'tsov, "Improvement of the Human Race," in *The Dawn of Human Genetics,* ed. V. V. Babkov (Cold Spring Harbor, NY: Cold Spring Harbor Laboratory Press, 2013), 86. Previously published in *Russkii evgenicheskii zhurnal,* no. 1, 1922.

2. Ibid., 70.

3. Ibid., 71.

4. Ibid., 83.

5. V. V. Bunak, "Novye dannye k voprosu o voine, kak biologicheskom faktore," *Russkii evgenicheskii zhurnal* 1, no. 2 (1923): 231.

6. Kol'tsov, "Improvement of the Human Race," 67.

7. V. V. Bunak, "Materialy dlia sravnitel'noi kharakteristiki sanitarnoi konstitutov evreev," *Russkii evgenicheskii zhurnal* 2–3 (1924): 142–152.

8. See, for example, Kol'tsov's generally positive review of E. Baur, E. Fischer, and F. Lenz, "Grundriss der menschlichen Erblichkeitslehre und Rassenhygiene," *Russkii evgenicheskii zhurnal,* no. 2 (1924): 2–3.

9. For an interesting discussion of eugenics issues, see Diane B. Paul, *The Politics of Heredity: Essays on Eugenics, Biomedicine, and the Nature-Nurture Debate* (Albany: State University of New York Press, 1998).

10. "Evgenicheskaia sterilizatsiia v Germanii," *Russkii evgenicheskii zhurnal* 3, no. 1 (1925).

11. A. S. Serebrovsky, "Anthropogenetics and Eugenics in a Socialist Society," in Babkov, *Dawn of Human Genetics,* 505–518.

12. Ibid., 511.

13. Ibid., 513.

14. Ibid., 515.

15. Ibid., 515–516.

16. S. N. Davidenkov, "Our Eugenic Prospects," in Babkov, *Dawn of Human Genetics,* 48–56.

17. "Letter from Muller to Stalin," in Babkov, *Dawn of Human Genetics,* 643–646.

18. Kammerer, *Inheritance of Acquired Characteristics,* 364.

19. Ibid.

20. A. E. Gaissinovitch, "The Origins of Soviet Genetics and the Struggle with Lamarckism, 1922–1929," trans. Mark B. Adams, *Journal of the History of Biology* 13 (Spring 1980): 15.

21. Given on November 21, 1924, published as "Evolutsionye teorii v biologii i marksizm," *Vestnik sovremennoi meditsiny* 9 (1925). See discussion in Gaissinovitch, "Origins of Soviet Genetics," 16.

22. Gaissinovitch, "Origins of Soviet Genetics."

23. Quoted in L. A. Blacher, *The Problem of the Inheritance of Acquired Characters* (New Delhi: Amerind, 1982), 130.

24. M. V. Volotskoi, *Fizicheskaia kul'tura s tochki zreniia evgeniki* (Moscow: Izdatel'stvo instituta fizicheskoi kul'tury imeni P. F. Lesgafta, 1924), 69.

25. T. H. Morgan and Iu. A. Filipchenko, *Nasledstvennye li priobretennye priznaki?* (Leningrad: Seiatel', 1925).

26. Quoted in N. M. Volotskoi, "Spornye voprosy evgeniki," *Vestnik kommunisticheskoi akademii* 20 (1927): 225.

27. For a discussion of the impact of Filipchenko's argument, see N. M. Volotskoi, "Spornye voprosy evgeniki," *Vestnik kommunisticheskoi akademii,* no. 20 (1927): 224–225.

28. Ibid.

29. Vasilii Slepkov, "Nasledstvennost' i otbor u cheloveka (Po povodu teoreticheskikh predposylok evgeniki)," *Pod znamenem marksizma,* no. 4 (1925): 102–122.

30. Ibid., 116.

31. Friedrich Engels, *Selected Works* (New York: International Publishers, 1950), 153.

32. Cheryl Logan even described Kammerer's dress as that of a "dandy." Cheryl A. Logan, *Hormones, Heredity, and Race: Spectacular Failure in Interwar Vienna* (New Brunswick, NJ: Rutgers University Press, 2013), Google eBook, 40.

33. *Chicago Sunday Tribune,* March 15, 1953, part 1, 18.

5. LYSENKO UP CLOSE

Parts of this chapter rely on my book *Moscow Stories* (Bloomington: Indiana University Press, 2006).

Epigraph: Conversation between Trofim Lysenko and Loren Graham, Moscow, 1971.

1. See Peter Pringle, *The Murder of Nikolai Vavilov: The Story of Stalin's Persecution of One of the Great Scientists of the Twentieth Century* (New York: Simon and Schuster, 2008).

2. Eduard L. Kolchinsky, "Nikolai Vavilov in the Years of Stalin's 'Revolution from Above' (1929–1932)" (unpublished manuscript given to Loren Graham, 2014).

3. Ibid., 19.

4. Immediately after this conversation with Lysenko, I wrote down careful notes of what each of us said. Also, see T. D. Lysenko, "Iarovizatsiia—eto milliony pudov dobavochnogo urozhaia," *Izvestiia,* February 15, 1935, 4.

5. Notes taken by Loren Graham after meeting Lysenko.

6. See the section on refuseniks in the *NOVA* program for which I was rapporteur: Martin Smith, "How Good Is Soviet Science?," *NOVA,* WGBH, 1987.

7. Notes taken by Loren Graham after meeting Lysenko. See also note 4.

8. Michael Gordin, "Lysenko Unemployed: Soviet Genetics after the Aftermath" (unpublished manuscript).

9. An investigation of Lysenko's dairy farm in 1965 by the Academy of Sciences revealed that he had been concealing his systematic elimination of poor milk producers and thereby fraudulently elevating his production statistics. See *Vestnik akademii nauk,* no. 11 (1965): especially 73, 91–92.

10. Valery N. Soyfer, *Lysenko and the Tragedy of Soviet Science* (New Brunswick, NJ: Rutgers University Press, 1994).

6. LYSENKO'S BIOLOGICAL VIEWS

Parts of this chapter rely on my book *Science and Philosophy in the Soviet Union* (New York: Alfred A. Knopf, 1972).

Epigraph: T. D. Lysenko, *Agrobiologiia* (Moscow: OGIZ, 1952), 221.

1. For a list of the most important of these publications, see Loren R. Graham, *Science and Philosophy in the Soviet Union* (New York: Alfred A. Knopf, 1972), 570–571.

2. See the discussion in Jan Sapp, *Beyond the Gene: Cytoplasmic Inheritance and the Struggle for Authority in Genetics* (Oxford: Oxford University Press, 1987). See especially p. 25.

3. T. D. Lysenko, *Izbrannye sochineniia,* vol. 2 (Moscow: OGIZ, 1958), 48.

4. An excellent discussion of this case is in P. S. Hudson and R. H. Richens, *The New Genetics in the Soviet Union* (Cambridge: School of Agriculture, 1946), 39.

5. See, for example, ibid., 32–51.

6. Lysenko, *Agrobiologiia,* 34–35 (emphasis in the original).

7. Nils Roll-Hansen, *The Lysenko Effect: The Politics of Science* (Amherst, NY: Humanity Books, 2005), 31–32.

8. For citations of much of the literature, see Richard Amasino, "Vernalization, Competence, and the Epigenetic Memory of Winter," *Plant Cell* 16 (October 2004): 2553–2559. This is a useful source, but unfortunately, it repeats the old canard that Lysenko aimed to produce future generations of improved citizens. Actually, Lysenko condemned any effort to apply his theories to human beings. Also, see O. N. Purvis, "The Physiological Analysis of Vernalization," in *Encyclopedia of Plant Physiology,* ed. W. H. Ruhland, vol. 16 (Berlin: Springer, 1961), 76–117.

9. John Evelyn, "Sylva, Or a Discourse on Forest Trees," Royal Society, October 16, 1662.

10. G. Gassner, "Beiträge zur physiologiischen Charakteristik sommer- und winter-annueller Gewächse, insbesondere der Getreidepflanzen," *Zeitschrift für Botanik* 10: 417–480.

11. "Vernalization," *McGraw Hill Encyclopedia of Science and Technology,* vol. 14 (New York, 1966), 305.

12. See his discussion in *Agrobiologiia,* especially p. 85.

13. For an example of this research, see Artem Loukoianov, Liuling Yan, Ann Blechl, Alexandra Sanchez, and Jorge Dubcovsky, "Regulation of *VRN-1*

Vernalization Genes in Normal and Transgenic Polyploid Wheat," *Plant Physiology* 138, no. 4 (August 2005): 2364–2373.

14. Lysenko, *Agrobiologiia*, 34.

15. Ibid., 83.

16. Lysenko, *Heredity and Its Variability,* trans. Th. Dobzhansky (New York: King's Crown Press, 1946). Lysenko on several occasions used the terms "genotype," "phenotype," and "genes," but in ways that showed his lack of understanding of what those terms mean to contemporary geneticists. See Nils Roll-Hansen, *The Lysenko Effect,* 165–167.

17. Lysenko, *Heredity and Its Variability.*

18. Ibid.

19. Lysenko, *Agrobiologiia,* 436.

20. A. A. Rukhkian, "Ob opisannom S. K. Karapetianom sluchae porozhdeniia leshchiny grabom," *Botanicheskii zhurnal* 38, no. 6 (1953): 885–891.

21. P. S. Hudson and R. H. Richens, *The New Genetics in the Soviet Union* (Cambridge: Cambridge University Press, 1946). The same scheme is given in T. D. Lysenko, *Heredity and Its Variability,* trans. Th. Dobzhansky (New York: Columbia University Press, 1946).

22. Lysenko, *Heredity and Its Variability,* 55–65.

23. Ibid., 51.

24. Lysenko, *Agrobiology,* 85. Lysenko quoted Michurin's belief that "the farther apart the crossed parental pairs are in respect to place of origin and environmental conditions, the more easily the hybrid seedlings adapt themselves to the environmental conditions of the new locality." If applied to humans, this view would be, of course, a strong argument for racial mixing. Lysenko did not extend his system to humans, here or elsewhere.

25. Hudson and Richens, *New Genetics in the Soviet Union,* 48.

26. Recent work on graft hybridization has shown that the exchange of genetic material between stock and scion can occur, but not in the way Lysenko proposed. Two researchers concluded in 2009: "Although our data demonstrate the exchange of genetic material between grafted plants, they do not lend support to the tenet of Lysenkoism that 'graft hybridization' would be

analogous to sexual hybridization. Instead, our finding that gene transfer is restricted to the contact zone between scion and stock indicates that the changes can become heritable only by lateral shoot formation from the graft site. However, there is some reported evidence for heritable alterations induced by grafting, and, in light of our findings, these cases certainly warrant detailed molecular investigation." See Sandra Stegemann and Ralph Bock, "Exchange of Genetic Material between Cells in Plant Tissue Grafts," *Science* 324, no. 5927 (May 1, 2009): 649–651.

27. T. D. Lysenko, *Izbrannye sochineniia,* vol. 2 (Moscow: Sel'khozgiz, 1958), 48.

28. Hudson and Richens, *New Genetics in the Soviet Union,* 42–43.

29. An example is John Langdon-Davies, who wrote that the controversy occurred because "a limit was being set to the extent to which environmental change at the hands of the U.S.S.R. planners could be expected to alter human nature permanently for the better." See Langdon-Davies, *Russia Puts the Clock Back: A Study of Soviet Science and Some British Scientists* (London: Gollancz, 1949), 58–59.

30. The knife cuts both ways. If one accepts the inheritance of acquired characteristics, building a "new man" may seem more possible, but the same argument would lead to racist positions or even belief in the superiority of aristocracies. As Julian Huxley wrote: "It is very fortunate for the human species that acquired characters are not readily impressed on the hereditary constitution. For if they were, the conditions of dirt, disease, and malnutrition in which the majority of mankind have lived for thousands of years would have produced a disastrous effect upon the race." See Huxley, *Heredity East and West (Lysenko and World Science)* (New York: H. Schuman, 1949), 138. Of course, any serious discussion of the implications for humans of a hypothetical inheritance of acquired characteristics would have to involve discussions of time: the number of generations required to fix the heredity of a new trait and the number of generations in new conditions to erase it.

31. *Pravda,* December 14, 1958.

32. L. C. Dunn, *A Short History of Genetics* (New York: McGraw-Hill, 1965), x.

33. Since Darwin ascribed validity to both natural selection and the inheritance of acquired characteristics, both the neo-Mendelians and Michurinists could call themselves Darwinists.

34. Lysenko, *Agrobiologiia,* 221.

7. EPIGENETICS

Epigraph: Nessa Carey, *The Epigenetics Revolution: How Modern Biology Is Rewriting Our Understanding of Genetics, Disease, and Inheritance* (London: ICON, 2012), 312.

1. Emma Young, "Rewriting Darwin: The New Non-Genetic Inheritance," *New Scientist,* July 9, 2008, 28–33.

2. Richard Dawkins, *The Selfish Gene* (New York: Oxford University Press, 1976). Dawkins has come under criticism because of the advent of epigenetics, but he remains very skeptical of the field. On August 21, 2011, he wrote, "I am heartily sick of the epigenetics bandwagon." See the Richard Dawkins Foundation website at https://richarddawkins.net.

3. Joshua Lederberg, "Cell Genetics and Hereditary Symbiosis," *Physiological Review* 32, no. 4 (1952): 403–430.

4. Barbara McClintock, "The Origin and Behavior of Mutable Loci in Maize," *Proceedings of the National Academy of Sciences of the United States of America* 36, no. 6 (1950): 344–355; "Introduction of Instability at Selected Loci in Maize," *Genetics* 38, no. 6 (1953): 579–599; "Some Parallels between Gene Control Systems in Maize and in Bacteria," *American Naturalist* 95 (September–October 1961): 265–277. See also Evelyn Fox Keller, *A Feeling for the Organism: The Life and Work of Barbara McClintock* (San Francisco: W. H. Freeman, 1983); and Nathaniel C. Comfort, *The Tangled Field: Barbara McClintock's Search for the Patterns of Genetic Control* (Cambridge, MA: Harvard University Press, 2001).

5. D. Nanney, "Epigenetic Control Systems," *Proceedings of the National Academy of Sciences* 44 (1958): 712–717.

6. See François Jacob and Jacques Monod, "Genetic Regulatory Mechanisms in the Synthesis of Proteins," *Journal of Molecular Biology,* no. 3 (1961): 318–356.

7. Otto E. Landman, "The Inheritance of Acquired Characteristics," *Annual Review of Genetics* 25 (December 1991): 1–20.

8. See J. Sapp, *Beyond the Gene: Cytoplasmic Inheritance and the Struggle for Authority in Genetics* (New York: Oxford University Press, 1987); Landman, "Inheritance of Acquired Characteristics," 1–20.

9. "Epigenetics" in the modern sense is a word coined by C. H. Waddington in 1942. However, there is still some controversy over its definition. A narrow definition is "heritable changes in the phenotype not derivable from sequence changes in the genotype." In this book this definition is preferred. However, there is another broader definition that "captures the variable effects of interactions between genes, embryonic development, and the environment." A vast literature is developing within this broader definition. See Daniel E. Lieberman, "Epigenetic Integration, Complexity, and Evolvability of the Head," in *Epigenetics: Linking Genotype and Phenotype in Development and Evolution,* ed. Benedikt Hallgrimsson and Brian K. Hall (Berkeley: University of California Press, 2011), 271–289.

10. R. B. Khesin, *Nepostoianstvo genoma* (Moscow: Nauka, 1984). For a discussion of the importance of this work, see M. D. Golubovsky, "Genome Inconstancy by Roman B. Khesin in Terms of the Conceptual History of Genetics," *Molecular Biology* 36, no. 2 (2002): 259–266. Golubovsky has been very helpful to the author in investigating genetics and epigenetics. See his works *Vek genetiki: Evoliutsiia idei i poniatii* (St. Petersburg: Borey Art, 2000); "Stanovlenie genetiki cheloveka," *Priroda,* no. 10 (2012): 53–63; "The Unity of the Whole and Freedom of Parts: Facultativeness Principle in the Hereditary System," *Vavilovskii zhurnal genetiki i selektsii* 15, no. 2 (2011): 423–431; and together with Kenneth G. Manton, "Three-Generation Approach in Biodemography Is Based on the Developmental Profiles and the Epigenetics of Female Gametes," *Frontiers in Bioscience* 10 (January 1, 2005): 187–191.

11. Edith Heard and Robert A. I. Martienssen, "Transgenerational Epigenetic Inheritance: Myths and Mechanisms," *Cell* 157 (March 27, 2014): 103.

12. Ibid., 95.

13. "Der Sieg über die Gene," *Der Spiegel* 32, August 9, 2010.

14. Martha Susiarjo and Marisa S. Bartolomei, "You Are What You Eat, But What about Your DNA? Parental Nutrition Influences the Health of Subsequent Generations through Epigenetic Changes in Germ Cells," *Science* 345 (August 15, 2014): 9–10.

15. Frances A. Champagne and Michael J. Meaney, "Transgenerational Effects of Social Environment on Variations in Maternal Care and Behav-

ioral Response to Novelty," *Behavioral Neuroscience* 121, no. 6 (2007): 1353–1363.

16. See the discussion in Lizzie Buchen, "Neuroscience: In Their Nurture," *Nature* 467 (September 8, 2010): 146–148, http://www.nature.com/news /2010/100908/full/467146a.html.

17. Michael J. Meaney et al., "Epigenetic Programming by Maternal Behavior," *Nature Neuroscience* 7 (2004): 847–859.

18. Dan Hurley, "Grandma's Experiences Leave a Mark on Your Genes," *Discover Magazine,* May 2013.

19. D. D. Francis et al., "Maternal Care, Gene Expression, and the Development of Individual Differences in Stress Reactivity," *Annals of New York Academy of Science* 896 (1999): 66–84; M. Szyf et al., "Maternal Programming of Steroid Receptor Expression and Phenotype through DNA Methylation in the Rat," *Frontiers in Neuroendocrinology* 26, nos. 3–4 (October–December 2005): 139–162.

20. P. O. McGowan et al., "Epigenetic Regulation of the Glucocorticoid Receptor in Human Brain Associates with Child Abuse," *Nature Neuroscience* 12 (2009): 342–348.

21. Buchen, "Neuroscience: In Their Nurture"; S. G. Gregory et al., "Genomic and Epigenetic Evidence for Oxytocin Receptor Deficiency in Autism," *BMC Medicine* 7, no. 62 (2009).

22. For example, see Marija Kundakovic and Frances A. Champagne, "Early-Life Experience, Epigenetics, and the Developing Brain," *Neuropsychopharmacology,* July 30, 2014, http://www.ncbi.nih.gov/pubmed/24917200.

23. Richard C. Francis, *Epigenetics: How Environment Shapes Our Genes* (New York: W. W. Norton, 2011), 1–5; Carey, *Epigenetics Revolution,* 101–104.

24. L. H. Lumey et al., "Cohort Profile: The Dutch Hunger Winter Families Study," *International Journal of Epidemiology* 36, no. 6 (2007): 1194–1204. See also Chris Bell, "Epigenetics: How to Alter Your Genes," *Telegraph,* October 16, 2013.

25. Francis, *Epigenetics: How Environment Shapes Our Genes,* 1; Carey, *Epigenetics Revolution,* 101–104.

8. THE RECENT REBIRTH OF LYSENKOISM IN RUSSIA

Epigraph: From *Biomolekula,* an online Russian biology newsletter, Biomo lecula.ru/content/1377.

1. Leonid Korochkin, "Neo-Lysenkoism in Russian Consciousness," http://www .lghtz.ru/archives/lg092002/Tetrad/art11_1.htm.

2. I. A. Zakharov-Gezekhus, "Eksgumatsiia lysenkovshchiny," http://www .plantgen.com/ru/genetika/storiye-genetiki/179-ekzgumacaiya-lysen kovshhiny.html, 31.01.2011.

3. Maksim Kalashnikov, http://m-kalashnikov.livejournal.com/1510946.html.

4. http://contrtv.ru.

5. http://zavtra.ru/content/archiv.

6. http://colonelcassad.livejournal.com.

7. http://biblioteka-dzvon.narod.ru/.

8. Kirill V. Zhivotovsky, http://kvzh.livejournal.com.

9. See, for example, the discussions of the Moscow University Alumni Club, www.moscowuniversityclub.ru; the discussions in the Novosibirsk University forum, http://forum.ngs.ru; the forum of Andrei Kuraev, http://kuraev.ru/smf/index.php?topic=288407.0; and the forum of S. Kara-Murza, "Vetka: Otkrytiia Lysenko Podtverzhdeny," vif2ne.ru:2009/nvz/forum /arhprint/305542.

10. See N. V. Ochinnikov et al., *Trofim Denisovich Lysenko: Sovetskii agronom, biolog, selektsioner* (Moscow: Samoobrazovanie, 2008), 31.

11. For example, I. Knuniants and L. Zubkov, "Shkoly v nauke," *Literaturnaia gazeta,* January 11, 1955, 1.

12. L. Korochkin, "Vo vlasti nevezhestva, Neolysenkovshchina v rossiiskom soznanii," *Literaturnaia gazeta,* March 6, 2002.

13. Mikhail Anokhin, "Akademik Lysenko i bednaia ovechka Dolli," *Literaturnaia gazeta,* March 18, 2009, 12.

14. Ibid.

15. See "Diskussiia v 'LG,' " http://lysenkoism.narod.ru/lgz-dv-gubarev.htm.

16. "Dva otklika na vystuplenie professor Anokhina," *Literaturnaia gazeta,* no. 23, June 3, 2009.

17. Mikhail Anokhin, "Nakormivshie lozh'iu," *Literaturnaia gazeta,* no. 5, 2015, 10.

18. Vladimir Gubarev, "D'iavol iz proshlogo," *Delovoi vtornik,* April 13, 2009.

19. "Lysenko v ovech'ei shkure," *Novaia gazeta,* no. 33, April 1, 2009.

20. Ibid.

21. Iurii Mukhin, *Prodazhnaia devka Genetika* (Moscow: Izdatel'stvo Bystrov, 2006).

22. Ibid., 47.

23. Ibid., 124.

24. Ibid., 77, 85, 169.

25. Ibid., 72–73.

26. Eduard L. Kolchinsky, "Current Attempts at Exoneration of 'Lysenkoism' and Their Causes" (unpublished paper), 4–6. I am grateful to Kolchinsky for help on this topic.

27. Ibid., 6.

28. Mukhin, *Prodazhnaia devka Genetika,* 44.

29. Ibid., 161.

30. Ibid., 42, 44.

31. Ibid., 108, 153, 154.

32. See Mukhin, "Rubbish," in *Prodazhnaia devka Genetika,* 119; Mukhin, "Obscurantism," in *Prodazhnaia devka Genetika,* 137; Mukhin, "Pride of Stupid Idiots," in *Prodazhnaia devka Genetika,* 146, 151.

33. Iurii Mukhin, "Ofitsial'noe priznanie zaslug T. D. Lysenko," http://www.ymuhin.ru/node/1048.

34. The impact factor of an academic journal is a measure of the average number of citations to articles published in the journal.

35. Edith Heard and Robert A. Martienssen, "Transgenerational Epigenetic Inheritance: Myths and Mechanisms," *Cell* 157 (March 27, 2014): 95–109.

36. Ibid., 95.

37. http://www.ymuhin.ru/node/1048.

38. "Trofim—ty prav!," *Zavtra,* April 1, 2014.

39. V. I. Pyzhenkov, *Nikolai Ivanovich Vavilov—Botanik, akademik, grazhdanin mira* (Moscow: Samoobrazovanie, 2009).

40. Ibid., especially 109–128.

41. Yevgeniia Albats, "Genius and the Villains," *Moscow News* 50 (December 13, 1987), 10.

42. S. Mironin, "Trofim Denisovich Lysenko," *Istoriia i sovremennost'*, January 11, 2013, http://mordikov.fatal,ru/lisenko.html.

43. Sigizmund Mironin, *Delo genetikov* (Moscow: Algoritm, 2008), 189–190.

44. P. F. Kononkov, *Dva mira—dve ideologii: O polozhenii v biologicheskikh i sel'skokhoziaistvennykh naukakh v Rossii v sovetskii i postsovetskii period* (Moscow: Luch, 2014).

45. Eduard Kolchinsky, e-mail to the author (and others), November 21, 2015.

46. Lev Zhivotovskii, *Neizvestnyi Lysenko* (Moscow: T-vo nauchnykh izdanii KMK, 2014), 47.

47. A. E. Murneek et al., eds., *Vernalization and Photoperiodism: A Symposium* (Waltham, MA: Chronica Botanica, 1948).

48. Ibid., 27.

49. Ibid., 37.

50. A. I. Shatalkin, *Reliatsionnye kontseptsii nasledstvennosti i bor'ba vokrug nikh v XX stoletii* (Moscow: T-vo nauchnykh izdanii KNK, 2015).

51. Ibid., 401.

9. SURPRISING EFFECTS OF THE NEW LYSENKOISM

Epigraph: V. S. Baranov, interview by Loren Graham, June 10, 2014.

1. M. D. Golubovsky, "Nekanonicheskie nasledstvennye izmeneniia," *Priroda*, no. 9 (2011): 9.

2. "To the 70th Anniversary of S. G. Inge-Vechtomov," *Russian Journal of Genetics* 45 (July 2009): 881–883. I know Inge-Vechtomov personally and have interviewed him several times.

3. See, for example, Tessa Roseboom, Susanne de Rooij, and Rebecca Painter, "The Dutch Famine and Its Long-Term Consequences for Adult Health," *Early Human Development* 82 (2006): 485–491.

4. Quoted in David Epstein, "How an 1836 Famine Altered the Genes of Children Born Decades Later," http://io9.com/how-an-1836 famine-altered -the-genes-of-children-born-d-1200001177, from his book *The Sports Gene: Inside the Science of Extraordinary Athletic Performance* (New York: Current, 2013).

5. Lars Olov Bygren, Gunnar Kaati, and Sören Edvinsson, "Longevity Determined by Paternal Ancestors' Nutrition during Their Slow Growth Period," *Acta Biotheretica* 49 (2001): 53–59; Marcus E. Pembrey et al., "Sex-Specific Male-Line Transgenerational Responses in Humans," *European Journal of Human Genetics* 14 (2006): 159–166.

6. Lidiia P. Khoroshinina, "Osobennosti somaticheskoi patologii u liudei starshikh vozrastnykh grupp, perezhivshikh v detstve blokadu Leningrada" (PhD diss., St. Petersburg Medical Academy, 2002). See also John Barber and Andrei Dzeniskovich, *Life and Death in Besieged Leningrad* (London: Palgrave Macmillan, 2005).

7. V. Bartel', "90-letiiu Akademika D. K. Beliaeva posviashchaetsia," *Nauka v Sibiri,* no. 34 (September 6, 2007).

8. See Arkady L. Merkel and Lyudmila N. Trut, "Behavior, Stress, and Evolution in Light of the Novosibirsk Selection Experiment," in *Transformations of Lamarckism,* ed. Snait B. Gissis and Eva Jablonka (Cambridge, MA: MIT Press, 2011), 171–180.

9. I am grateful to Eva Jablonka and Marion Lamb for describing these events and for helping me on this topic. Eva Jablonka, e-mail message to author, January 31, 2014.

10. S. Iu. Vert'yanov, "Great Damage," *Obshchaia Biologiia* (2012): 198.

11. N. V. Ovchinnikov, *Akademik Trofim Denisovich Lysenko* (Moscow: Luch, 2010), 85–87.

12. N. V. Ovchinnikov, Arkhiv RAN, f. 1521, op. 1, no. 281, 87.

13. P. F. Kononkov, *Dva mira—dve ideologii. O polozhenii v biologicheskikh i sel'skokhoziaistvennykh naukakh v Rossii v sovetskii i postsovetskii period* (Moscow: Luch, 2014); N. V. Ovchinnikov et al., *Trofim Denisovich Lysenko—Sovetskii agronom, biolog, selektsioner* (Moscow: Samoobrazovanie, 2008).

14. William R. Rice, Urban Friberg, and Sergey Gavrilets, "Homosexuality as a Consequence of Epigenetically Canalized Sexual Development," *Quarterly Review of Biology* 87 (December 2012): 343–368.

15. CNN, December 12, 2012; _Science,_ December 11, 2012; _RIA Novosti,_ December 12, 2012; _Komsomolskaia Pravda,_ December 12, 2012.

16. Rice, Freiberg, and Gavrilets, "Homosexuality," 358.

17. Brian G. Dias and Kerry J. Ressler, "Parental Olfactory Experience Influences Behavior and Neural Structure in Subsequent Generations," _Nature Neuroscience_ 17 (2014): 89–96; http://www.nature.com/neuro/journal/v17ul/full/nn.3594.html.

18. Vladimir Kozlov, "Strakh peredaetsia po nasledstvu?," _Gorodskie novosti,_ July 25, 2013, http://www.gornovosti.ru/tema/neotlozhka/strakh-peredayetsya-po-nasledstvu40779.htm.

19. Oleg Kolosov, "Strakh peredaetsia russkim geneticheski rezul'tate bol'shevitskikh repressii?," December 30, 2013, http://rons-inform.livejournal.com/1303024.html.

10. ANTI-LYSENKO RUSSIAN SUPPORTERS OF THE INHERITANCE OF ACQUIRED CHARACTERISTICS

Epigraph: A. A. Liubishchev, _V zashchitu nauki_ (Leningrad: Nauka, 1991), 35–36.

1. L. I. Blacher, _The Problem of the Inheritance of Acquired Characters: A History of A Priori and Empirical Methods Used to Find a Solution_ (New Delhi: Amerind, 1983). Originally published as L. Ia. Bliakher, _Problema nasledovaniia priobretennykh priznakov: Istoriia apriornykh i empiricheskikh popytok ee resheniia_ (Moscow: Nauka, 1971).

2. Some of them were published much later, after the collapse of the Soviet Union, in M. D. Golubovsky, ed., _V zashchitu nauki_ (Leningrad, USSR: Nauka, 1991).

3. Ibid., 46–47, 54–56.

4. Ibid., 23.

5. Ibid., 20, 26.

6. Ibid., 26.

7. Ibid., 248–272.

8. Golubovsky, _V zashchitu nauki,_ 260–272.

9. A. A. Liubishchev, "O dvukh stat'iakh po genetike," in Golubovsky, _V zashchitu nauki,_ 248–273. I am grateful to Michael Golubovsky for drawing my

attention to this article. See also A. A. Liubishchev, *O monopolii T. D. Lysenko v biologii* (Ulyanovsk, Russia: UGPU, 2004).

10. Golubovsky, *V zashchitu nauki,* 270–271.

CONCLUSION

1. For a discussion of the importance of molecular biology in the prehistory of epigenetics, see references to the work of Lederberg, Nanney, McClintock, Lwoff, Jacob, and Monod.

2. I am indebted to Michael Meaney for this insight.

3. Lysenko to N. P. Dubinin, September 25, 1974, in Eduard L. Kolchinsky, "Current Attempts at Exonerating 'Lysenkoism' and Their Causes" (unpublished paper), 9.

4. C. D. Darlington, "The Retreat from Science in Soviet Russia," in *Death of a Science in Russia,* ed. Conway Zirkle (Philadelphia: University of Pennsylvania Press, 1949), 72–73.

5. Carole Lartigue et al., "Genome Transplantation in Bacteria: Changing One Species to Another," *Science* 317 (August 3, 2007): 632–638.

6. See the discussion in Jan Sapp, *Beyond the Gene: Cytoplasmic Inheritance and the Struggle for Authority in Genetics* (New York: Oxford University Press, 1987), 179.

7. There were a few exceptions, most notably nature protection societies, such as the Moscow Society of Naturalists and the All-Russian Society for the Protection of Nature, which Douglas Weiner has called "a little corner of freedom." See Weiner, *A Little Corner of Freedom: Russian Nature Protection from Stalin to Gorbachev* (Berkeley: University of California Press, 1999).

8. Florian Maderspacher, "Lysenko Rising," *Current Biology* 20, no. 19 (2010): R835–R837. The respected researchers Edith Heard and Robert Martienssen also contributed to this trend by inaccurately attributing to Lysenko the "discovery" of cold-induced vernalization. Edith Heard and Robert A. Martienssen, "Transgenerational Epigenetic Inheritance: Myths and Mechanisms," *Cell* 157 (March 27, 2014): 95–109.

BIBLIOGRAPHY

Adams, Mark B. "The Founding of Population Genetics: Contributions of the Chetverikov School 1924–1934." *Journal of the History of Biology* (Spring 1968): 23–39.

———. "Genetics and the Soviet Scientific Community, 1948–1965." PhD diss., Harvard University, 1972.

———. "Science, Ideology, and Structure: The Kol'tsov Institute, 1900–1970." In *The Social Context of Soviet Science,* edited by Linda L. Lubrano and Susan Gross Solomon, 173–204. Boulder, CO: Westview Press, 1980.

———, ed. *The Wellborn Science: Eugenics in Germany, France, Brazil, and Russia.* New York: Oxford University Press, 1990.

Agol, I. I. "Dialektika i metafizika v biologii." *Pod znamenem marksizma,* no. 3 (1926): 142–159.

———. *Khochu zhit': Povest'.* Moscow: Khudozhestvennaia literatura, 1936.

Agranovskii, Anatolii. "Nauka na veru ne prinimaet." *Literaturnaia gazeta,* January 23, 1965, 2.

Albats, Yevgeniia. "Genius and the Villains." *Moscow News,* December 13, 1987, 10.

Allen, G. *Thomas Hunt Morgan: The Man and His Science.* Princeton, NJ: Princeton University Press, 1978.

Allis, C. D., T. Jenuwein, D. Reinberg, and M. L. Caparros, eds. *Epigenetics.* Cold Spring Harbor, NY: Cold Spring Harbor Laboratory Press, 2007.

Amasino, Richard. "Vernalization, Competence, and the Epigenetic Memory of Winter." *Plant Cell* 16 (October 2004): 253–259.

Anokhin, Mikhail. "Akademik Lysenko i bednaia ovechka Dolli." *Literaturnaia gazeta,* March 18, 2009, 12.

Anway, M. D., and M. K. Skinner. "Epigenetic Programming of the Germ Line: Effects of Endocrine Disruptors on the Development of Transgenerational Disease." *Reproductive Medicine Online* 16, no. 1 (2008): 23–25.

Babkov, V. V., ed. *The Dawn of Human Genetics.* Cold Spring Harbor, NY: Cold Spring Harbor Laboratory Press, 2013.

Barber, John, and Andrei Dzeniskovic. *Life and Death in Besieged Leningrad.* London: Palgrave Macmillan, 2005.

Bartel', V. "90-letiiu akademika D. K. Beliaeva posviashchaetsia." *Nauka v Sibiri* 34 (September 6, 2007): 3–7.

Bateson, William. "Dr. Kammerer's *Alytes.*" *Nature* 111 (1923): 738–739.

Baur, E., E. Fischer, and F. Lenz. *Grundriss der menschlichen Erblichkeitslehre und Rassenhygiene.* Munich: Lehmann, 1921.

Baylin, Stephen B., and Peter A. Jones. "A Decade of Exploring the Cancer Epigenome—Biological and Translational Implications." *Nature Reviews Cancer* 11 (October 2011): 726–734.

Beadle, George, and Muriel Beadle. *The Language of Life: An Introduction to the Science of Genetics.* Garden City, NY: Doubleday, 1966.

Beck, S. A., et al. "From Genomics to Epigenomics: A Loftier View of Life." *Nature Biotechnology* 11, no. 12 (1999): 1144.

Beisson, J., and T. M. Sonneborn. "Cytoplasmic Inheritance of the Organization of the Cell Cortex in *Paramecium aurelia.*" *Proceedings of the National Academy of Sciences USA* 53, no. 2 (1965): 275–282.

Bell, Chris. "Epigenetics: How to Alter Your Genes." *Telegraph,* October 16, 2013. http://www. telegraph.com.

Belyaev, D. K., A. O. Ruvinsky, and L. N. Trut. "Inherited Activation-Inactivation of the Star Gene in Foxes: Its Bearing on the Problem of Domestication." *Journal of Heredity* 72, no. 4 (1981): 267–274.

Berg, R. L., and N. V. Timofeev-Ressovskii. "Paths of Evolution of the Genotype." *Problems of Cybernetics* 10, no. 5 (1961): 292. (Joint Publication Research Service, 10, 292, January 1962.)

Berg, Raisa. *Sukhovie: Vospominaniia genetika.* Moscow: Pamiatniki istoricheskoi mysli, 2003.

Bernal, J. D. "The Biological Controversy in the Soviet Union and Its Implications." *Modern Quarterly* 4, no. 3 (1949): 203–217.

Berz, Peter, and Klaus Taschwer. "Afterword." *Arthur Koestler, Der Krötenküsser: Der Fall des Biologen Paul Kammerer.* Vienna: Czernin, 2010.

Bestor, T. H. "DNA Methylation: Evolution of a Bacterial Immune Function into a Regulator of Gene Expression and Genome Structure in Higher Eukaryotes." *Philosophical Transactions of the Royal Society of London,* ser. B: Biological Sciences, 326, no. 1235 (1990): 179–187.

Bestor, Timothy H., John R. Edwards, and Mathieu Bouland. "Notes on the Role of Dynamic DNA Methylation in Mammalian Development." *Proceedings of the National Academy of Sciences,* November 2014, 6796–6799.

Beurton, P. J., R. Falk, and H.-J. Rheinberger, eds. *The Concept of the Gene in Development and Evolution: Historical and Epistemological Perspective.* Cambridge: Cambridge University Press, 2000.

Biliya, S., and L. A. Bulla Jr. "Genomic Imprinting: The Influence of Differential Methylation in the Two Sexes." *Experimental Biology and Medicine* 235, no. 2 (2010): 139–147.

Bird, A. "CpG-Rich Islands and the Function of DNA Methylation." *Nature* 321 (May 15, 1986): 209–213.

———. "Perceptions of Epigenetics." *Nature* 447 (May 2007): 396–398.

Blacher, L. I. *The Problem of the Inheritance of Acquired Characters: A History of A Priori and Empirical Methods Used to Find a Solution.* Edited by F. B. Churchill. New Delhi: Amerind, 1982. Originally published as L. Ia. Bliakher, *Problema nasledovaniia priobretennykh priznakov: Istoriia apriornykh i empiricheskikh popytok ee resheniia.* Moscow: Nauka, 1971.

Blewitt, M. E., N. K. Vickaryous, A. Paldi, H. Koseki, and E. Whitelaw. "Dynamic Reprogramming of DNA Methylation at an Epigenetically Sensitive Allele in Mice." *PLoS Genetics* 2, no. 4 (2006): 399–405.

Bressan, Ray A., Jian-Kang Zhu, Michael J. Van Oosten, Hans J. Bohnert, Viswanathan Chinnusamy, and Albino Maggio. "Epigenetics Connects the Genome to Its Environment." In *Plant Breeding Reviews* 38, edited by Jules Janick (November 2014). doi: 10.1002/9781118916865.ch03.

Buchen, Lizzie. "Neuroscience: In Their Nurture." *Nature* 467 (September 8, 2010): 146–148.

Bunak, V. V. "Novye dannye k voprosu o voine, kak biologicheskom faktore." *Russkii evgenicheskii zhurnal,* no. 2 (1923): 223–232.

Burdge, G. C., S. P. Hoile, T. Uller, N. A. Thomas, P. D. Gluckman, et al. "Progressive, Transgenerational Changes in Offspring Phenotype and Epigenotype

Following Nutritional Transition." *PLoS ONE* 6, no. 11 (2011). http://journals .plos.org.

Burdge, G. C., J. Slater-Jefferies, C. Torrens, E. S. Phillips, M. A. Hanson, and K. A. Lillycrop. "Dietary Protein Restriction of Pregnant Rats in the F_0 Generation Induces Altered Methylation of Hepatic Gene Promoters in the Adult Male Offspring in the F_1 and F_2 Generations." *British Journal of Nutrition* 97, no. 3 (2007): 435–439.

Burkhardt, Richard W., Jr. *The Spirit of System: Lamarck and Evolutionary Biology.* Cambridge, MA: Harvard University Press, 1977.

Burua, S., and M. A. Junaid. "Lifestyle, Pregnancy and Epigenetic Effects." *Epigenomics* 1 (February 7): 85–102.

Bygren, Lars Olov, Gunnar Kaati, and Sören Edvinsson. "Longevity Determined by Paternal Ancestors' Nutrition during Their Slow Growth Period." *Acta Biotheretica* 49 (2001): 53–59.

Calatayud, F., and C. Belzung. "Emotional Reactivity in Mice, a Case of Nongenetic Heredity?" *Physiology and Behavior* 74, no. 3 (2001): 355–362.

Callinan, P. A., and A. P. Feinberg. "The Emerging Science of Epigenomics." Supplement 1, *Human Molecular Genetics* 15 (2006): R95–R101.

Carey, Nessa. *The Epigenetic Revolution: How Modern Biology Is Rewriting Our Understanding of Genetics, Disease, and Inheritance.* London: ICON, 2012.

Carlson, Elof Axel. *Genes, Radiation and Society: The Life and Work of H. J. Muller.* Ithaca, NY: Cornell University Press, 1981.

Champagne, F. A., and J. P. Curley. "Epigenetic Mechanisms Mediating the Long-Term Effects of Maternal Care on Development." *Neuroscience and Biobehavioral Reviews* 33, no. 4 (2009): 593–600.

Champagne, F. A., and M. J. Meaney. "Like Mother, Like Daughter: Evidence for Non-Genomic Transmission of Parental Behavior and Stress Responsivity." *Progress in Brain Research* 133 (2001): 287–302.

Champagne, F. A., et al. "Maternal Care Associated with Methylation of the Estrogen Receptor–Alpha1b Promoter and Estrogen Receptor–Alpha Expression in the Medial Preoptic Area of Female Offspring." *Endocrinology* 147 (June 2006): 2905–2915.

Champagne, Frances A., and Michael J. Meaney. "Transgenerational Effects of Social Environment on Variations in Maternal Care and Behavioral Response to Novelty." *Behavioral Neuroscience* 121, no. 6 (2007): 1353–1363.

Chong, S., N. Vickaryous, A. Ashe, N. Zamudio, N. Youngson, S. Hemley, et al. "Modifiers of Epigenetic Reprogramming Show Paternal Effects in the Mouse." *Nature Genetics* 39, no. 5 (2007): 614–622.

Chong, S., and E. Whitelaw. "Epigenetic Germline Inheritance." *Current Opinion in Genetics and Development* 14 (2004): 692–696.

Churchill, Frederick B. *August Weismann: Development, Heredity, and Evolution.* Cambridge, MA: Harvard University Press, 2015.

Comfort, Nathaniel. *The Tangled Field: Barbara McClintock's Search for the Patterns of Gene Control.* Cambridge, MA: Harvard University Press, 2003.

Crews, D., A. C. Gore, T. S. Hsu, N. L. Dangleben, M. Spinetta, T. Schallert, M. D. Anway, and M. K. Skinner. "Transgenerational Epigenetic Imprints on Mate Preference." *Proceedings of the National Academy of Sciences USA* 104, no. 14 (2007): 5942–5946.

Darlington, C. D. "The Retreat from Science in Soviet Russia." In *Death of a Science in Russia,* edited by Conway Zirkle, 67–80. Philadelphia: University of Pennsylvania Press, 1949.

Davidenkov, S. N. "Our Eugenic Prospects." In *The Dawn of Human Genetics,* edited by V. V. Babkov, 48–56. Cold Spring Harbor, NY: Cold Spring Harbor Laboratory Press, 2013.

Dawkins, Richard. *The Selfish Gene.* New York: Oxford University Press, 1976.

deJong-Lambert, William. *The Cold War Politics of Genetic Research: An Introduction to the Lysenko Affair.* Dordrecht, Netherlands: Springer, 2012.

———. "Hermann J. Muller, Theodosius Dobzhansky, Leslie Clarence Dunn and the Reaction to Lysenkoism in the United States." *Journal of Cold War Studies* 15 (Winter 2013): 78–118.

deJong-Lambert, William, and Nikolai Krementsev. "On Labels and Issues: The Lysenko Controversy and the Cold War." *Journal of the History of Biology* 45, no. 1 (2012): 373–388.

Delage, B., and Dashwood, R. H. "Dietary Manipulation of Histone Structure and Function." *Annual Review of Nutrition* 28 (2008): 347–366.

Denenberg, V. H., and K. M. Rosenberg. "Nongenetic Transmission of Information." *Nature* 216, no. 5115 (1967): 549–550.

Dias, Brian G., and Kerry J. Ressler. "Parental Olfactory Experience Influences Behavior and Neural Structure in Subsequent Generations." *Nature Neuroscience* 17 (2014): 89–96.

Dobrzhanskii, F. G. *Chto i kak nasleduetsia u zhivykh sushchestv?* Leningrad, USSR: Gosudarstvennoe izdatel'stvo, 1926, 51–64.

Dobzhansky, Theodosius. *The Biological Basis of Human Freedom*. New York: Columbia University Press, 1956.

———. *Genetics and the Origin of Species*. New York: Columbia University Press, 1937.

———. "N. I. Vavilov, a Martyr of Genetics, 1887–1942." *Journal of Heredity* 38 (August 1947): 229–230.

Dubinin, N. P. "Filosofskie i sotsiologicheskie aspekty genetiki cheloveka." *Voprosy filosofii,* no. 1 (1971): 36–45.

———. "I. V. Michurin i sovremennaia genetika." *Voprosy filosofii,* no. 6 (1966): 59–70.

———. *Istoriia i tragediia sovetskoi genetiki*. Moscow: Nauka, 1992.

———. *Izbrannye Trudy*. Moscow: Nauka, 2000.

———. *Vechnoe dvizhenie*. Moscow: Politizdat, 1989.

Duchinskii, F. "Darvinizm, lamarkizm i neodarvinizm." *Pod znamenem marksizma,* nos. 7–8 (1926): 95–122.

Dunn, L. C. *A Short History of Genetics*. New York: McGraw-Hill, 1965.

Evelyn, John. "Sylva, Or a Discourse on Forest Trees." Paper presented at the Royal Society, October 16, 1662.

Filipchenko, Iu. A. "Spornye voprosy evgeniki." *Vestnik kommunisticheskoi akademii,* no. 20 (1927): 212–254.

Fish, E. W., D. Shahrokh, R. Bagot, C. Caldji, T. Bredy, M. Szyf, and M. J. Meaney. "Epigenetic Programming of Stress Responses through Variation in Maternal Care." *Annals of the New York Academy of Science* 1036 (2004): 167–180.

Fitzpatrick, Sheila. *The Commissariat of Enlightenment: Soviet Organization of Education and the Arts under Lunacharsky.* Cambridge: Cambridge University Press, 1970.

Flanagan, J. M., V. Popendikyte, N. Pozdniakovaite, M. Sobolev, A. Assadzadeh, A. Schumacher, M. Zangeneh, et al. "Intra- and Interindividual Epigenetic Variation in Human Germ Cells." *American Journal of Human Genetics* 79, no. 1 (2006): 67–84.

Flintoft, L., ed. "Focus On: Epigenetics." *Nature Reviews Genetics* 8, no. 4 (2007): 245–314.

Francis, D. D., F. A. Champagne, et al. "Maternal Care, Gene Expression, and the Development of Individual Differences in Stress Reactivity." *Annals of New York Academy of Science* 896 (1999): 66–84.

Francis, D. D., and M. J. Meaney. "Maternal Care and the Development of Stress Responses." *Current Opinion in Neurobiology* 9, no. 1 (1999): 128–134.

Francis, Richard C. *Epigenetics: How Environment Shapes Our Genes.* New York: W. W. Norton, 2011.

Frolov, I. T. *Genetika i dialektika.* Moscow: Nauka, 1968.

Fyfe, James. *Lysenko Is Right.* London: Lawrence and Wishart, 1950.

Gaissinovitch, A. E. "The Origins of Soviet Genetics and the Struggle with Lamarckism, 1922–1929." Translated by Mark B. Adams. *Journal of the History of Biology* 13 (Spring 1980): 1–51.

Gassner, G. "Beiträge zur physiologiischen Charakteristik sommer-und winterannueller Gewächse, insbesondere der Getreidepflanzen." *Zeitschrift für Botanik* 10 (1918): 417–480.

Genereux, D. P., B. E. Miner, C. T. Bergstrom, and C. D. Laird. "A Population-Epigenetic Model to Infer Site-Specific Methylation Rates from Double-Stranded DNA Methylation Patterns." *Proceedings of the National Academy of Sciences USA* 102, no. 16 (2005): 5802–5807.

Gillispie, Charles Coulston. "Lamarck and Darwin in the History of Science." In *Forerunners of Darwin: 1745–1859,* edited by Bentley Glass, Owsei Temkin, and William L. Straus Jr., 265–291. Baltimore: Johns Hopkins University Press, 1968.

Gissis, Snait B., and Eva Jablonka, eds. *Transformations of Lamarckism.* Cambridge, MA: MIT Press, 2011.

Gluckman, P. D., M. A. Hanson, and A. S. Beedle. "Non-Genomic Transgenerational Inheritance of Disease Risk." *BioEssays* 29, no. 2 (2007): 145–154.

Gokhman, D., et al. "Reconstructing the DNA Methylation Maps of the Neandertal and the Denisovan." *Science* 344 (2014): 523–527.

Goldschmidt, Richard B. *In and Out of the Ivory Tower: The Autobiography of Richard B. Goldschmidt.* Seattle: University of Washington Press, 1960.

Golubovsky, M. D. "Genome Inconstancy by Roman B. Khesin in Terms of the Conceptual History of Genetics." *Molecular Biology* 36, no. 2 (2002): 259–266.

———. "Nekanonicheskie nasledstvennye izmeneniia." *Priroda,* no. 9 (2011): 53–63.

———. "Stanovlenie genetiki cheloveka." *Priroda,* no. 10 (2012): 53–63.

———. "The Unity of the Whole and Freedom of Parts: Facultativeness Principle in the Hereditary System." *Vavilovskii zhurnal genetiki i selektsii* 15, no. 2: 423–431.

———. *Vek genetiki: Evoliutsiia idei i poniatii.* St. Petersburg, Russia: Borei Art, 2000.

———, ed. *V zashchitu nauki.* Leningrad, USSR: Nauka, 1991.

Gordin, Michael. "How Lysenkoism Became Pseudoscience: Dobzhansky to Velikovsky." *Journal of the History of Biology* 45, no. 3 (2012): 443–468.

———. "Lysenko Unemployed: Soviet Genetics after the Aftermath." Unpublished manuscript.

Gould, Stephen J. *The Structure of Evolutionary Theory.* Cambridge, MA: Belknap Press of Harvard University Press, 2002.

Graham, Loren R. "Eugenics: Weimar Germany and Soviet Russia." In *Between Science and Values,* 217–256. New York: Columbia University Press, 1981.

———. "Genetics." In *Science and Philosophy in the Soviet Union,* 195–256. New York: Alfred Knopf, 1972.

———. *Moscow Stories.* Bloomington: Indiana University Press, 2006.

———. "Reasons for Studying Soviet Science: The Example of Genetic Engineering." In *The Social Context of Soviet Science,* edited by Linda L. Lubrano and Susan Gross Solomon, 205–240. Boulder, CA: Westview Press, 1980.

———. "Science and Values: The Eugenics Movement in Germany and Russia in the 1920s." *American Historical Review* 82 (December 1977): 1133–1164.

Grant-Downton, R. T., and H. G. Dickinson. "Epigenetics and Its Implications for Plant Biology: The 'Epigenetic Epiphany': Epigenetics, Evolution, and Beyond." *Annals of Botany* 97, no. 1 (2006): 11–27.

Gregory, S. G., et al. "Genomic and Epigenetic Evidence for Oxytocin Receptor Deficiency in Autism." *BMC Medicine* 7, no. 62 (2009).

Gubarev, Vladimir. "D'iavol iz proshlogo." *Delovoi vtornik,* April 13, 2009. http://lysenkoism.narod.

Gurdon, J. B., T. R. Elsdale, and M. Fischberg. "Sexually Mature Individuals of *Xenopus laevis* from the Transplantation of Single-Somatic Nuclei." *Nature* 182 (July 5, 1958): 64–65.

Haig, D. "The (Dual) Origin of Epigenetics." *Cold Spring Harbor Symposia on Quantitative Biology* 69: 67–70.

Haldane, J. B. S. *The Inequality of Man.* London: Chatto and Windus, 1932.

Hale, Piers J. *Political Descent: Malthus, Mutualism, and the Politics of Evolution in Victorian England.* Chicago: University of Chicago Press, 2014.

Harman, Oren Solomon. "C. D. Darlington and the British and American Reaction to Lysenko and the Soviet Conception of Science." *Journal of the History of Biology* 36, no. 2 (2003): 309–352.

Heard, Edith, and Robert A. I. Martienssen. "Transgenerational Epigenetic Inheritance: Myths and Mechanisms." *Cell* 157 (March 27, 2014): 95–109.

Henderson, I. R., and Jacobsen, S. E. "Epigenetic Inheritance in Plants." *Nature* 447, no. 7143 (2007): 418–424.

Holliday, R. "Epigenetics: A Historical Overview." *Epigenetics* 1, no. 2 (2006): 76–80.

———. "Epigenetics Comes of Age in the Twenty-First Century." *Journal of Genetics* 81, no. 1 (2002): 1–4.

———. "The Significance of DNA Methylation in Cellular Aging." In *Molecular Biology of Aging,* edited by A. D. Woodhead et al., 269–283. New York: Plenum Press, 1984.

Howard, Walter L. "Luther Burbank: A Victim of Hero Worship." *Chronica Botanica* 9, nos. 5–6 (1945): 299–520.

Hudson, P. S., and R. H. Richens. *The New Genetics in the Soviet Union.* Cambridge: School of Agriculture, 1946.

Hunter, Philip. "What Genes Remember." May 2008. http://www.prospect-magazine.co.uk/article_details.php?id=10140.

Hurley, Dan. "Grandma's Experiences Leave a Mark on Your Genes." *Discover Magazine* (May 2013). http://discovermagazine.

Huxley, Julian. *Heredity East and West (Lysenko and World Science)*. London: H. Schuman, 1949.

———. *A Scientist among the Soviets*. London: Chatto and Windus, 1932.

Il'in, M. M. "Filogenez pokrytosemennykh s pozitsii michurinskoi biologii." *Botanicheskii zhurnal* 38, no. 1 (1953): 97–118.

Iogansen, Nil's. "Trofim Lysenko: Genii ili sharlatan?" *Kul'tura,* July 2, 2015.

Jablonka, Eva, and Marion J. Lamb. *Epigenetic Inheritance and Evolution: The Lamarckian Dimension*. Oxford: Oxford University Press, 1995.

———. "Epigenetic Inheritance as a Mediator of Genetic Changes during Evolution." Paper presented in Novosibirsk, 2007, compliments of Eva Jablonka.

———. *Evolution in Four Dimensions: Genetic, Epigenetic, Behavioral, and Symbolic Variation in the History of Life*. Cambridge, MA: MIT Press, 2005.

———. "The Inheritance of Acquired Epigenetic Variations." *Journal of Theoretical Biology* 139, no. 1 (1989): 69–83.

Jablonka, Eva, and Gal Raz. "Transgenerational Epigenetic Inheritance: Prevalence, Mechanisms, and Implications for the Study of Heredity and Evolution." *Quarterly Review of Biology* 84 (June 2009): 131–176.

Jacob, François, and Jacques Monod. "Genetic Regulatory Mechanisms in the Synthesis of Proteins." *Journal of Molecular Biology,* no. 3 (1961): 318–356.

Jirtle, R. L., and M. K. Skinner. "Environmental Epigenomics and Disease Susceptibility." *Nature Reviews Genetics* 8, no. 4 (2007): 253–262.

Jones, Bryony. "Epigenetics: Histones Pass the Message On." *Nature Reviews Genetics* 16, no. 3 (2015). doi: 10.1038/nrg3876.

Joravsky, David. "The First Stage of Michurinism." In *Essays in Russian and Soviet History,* edited by J. S. Curtiss, 120–132. New York: Columbia University Press, 1963.

———. *The Lysenko Affair*. Cambridge, MA: Harvard University Press, 1970.

————. "Soviet Marxism and Biology before Lysenko." *Journal of the History of Ideas* 20, no. 1 (1959): 85–104.

————. *Soviet Marxism and Natural Science, 1917–1932*. New York: Columbia University Press, 1961.

Jorgensen, R. A. "Restructuring the Genome in Response to Adaptive Challenge: McClintock's Bold Conjecture Revisited." *Cold Spring Harbor Symposia on Quantitative Biology* 69 (2004): 349–354.

Josephson, Paul. *New Atlantis Revisited: Akademgorodok, the Siberian City of Science*. Princeton, NJ: Princeton University Press, 1997.

Kaati, G., L. O. Bygren, M. Pembrey, and J. Sjöstrom. "Transgenerational Response to Nutrition, Early Life Circumstances and Longevity." *European Journal of Human Genetics* 15 (2007): 784–790.

Kammerer, Paul. "Breeding Experiments on the Inheritance of Acquired Characters." *Nature* 111 (1923): 637–640.

————. "Das Darwinmuseum zu Moskau." *Monistische Monatshefte* 11 (1926): 377–382.

————. *The Inheritance of Acquired Characteristics*. New York: Boni and Liveright, 1924.

Katz, L. A. "Genomes: Epigenomics and the Future of Genome Sciences." *Current Biology* 16 (2006): R996–R997.

Keller, Evelyn Fox. *The Century of the Gene*. Cambridge: Cambridge University Press, 2000.

————. *A Feeling for the Organism: The Life and Work of Barbara McClintock*. San Francisco: W. H. Freeman, 1983.

Kennedy, Scott. "Multigenerational Epigenetic Inheritance." Lecture presented at Harvard Medical School, NRB, Boston, MA, December 4, 2013.

Khesin, R. B. *Nepostoianstvo genoma*. Moscow: Nauka, 1984.

Khoroshinina, Lidiia. "Osobennosti somaticheskoi patologii u liudei starshikh vozrastnykh grupp, perezhivshikh v detstve blokadu Leningrada." PhD diss., St. Petersburg Medical Academy, 2002.

Kiklenka, Keith, et al. "Disruption of Histone Methylation in Developing Sperm Impairs Offspring Health Transgenerationally." *Science* 350, no. 6261 (November 6, 2015): DOI:10.1126/science.aab2006.

Knuniants, I., and L. Zubkov. "Shkoly v nauke." *Literaturnaia gazeta,* January 11, 1955, 1.

Koestler, Arthur. *The Case of the Midwife Toad: A Scientific Mystery Revisited.* New York: Random House, 1971.

Kojevnikov, Alexei B. *Stalin's Great Science: The Times and Adventures of Soviet Physicists.* London: Imperial College Press, 2004.

Kol', A. "Prikladnaia botanika ili leninskoe obnovlenie zemli." *Ekonomicheskaia zhizn',* January 29, 1931, 2.

Kolbanovskii, V. "Spornye voprosy genetiki i selektsii (obshchii obzor soveshchaniia)." *Pod znamenem marksizma,* no. 11 (1939): 95.

Kolchinsky, Eduard. *Biology in Germany and Russia-USSR: Under Conditions of Social-Political Crises of the First Half of the XX Century.* St. Petersburg, Russia: Nestor-Historia, 2007.

———. "Current Attempts at Exoneration of 'Lysenkoism' and Their Causes." Unpublished paper given to Loren Graham, 2014.

———. "Nikolai Vavilov in the Years of Stalin's 'Revolution from Above' (1929–1932)." Manuscript given to Loren Graham, 2014.

———. "N. I. Vavilov v prostranstve istoriko-nauchnykh diskusii." Paper given at XII Vavilovskoe chtenie, conference held at the Institute of General Genetics, Russian Academy of Sciences, November 18, 2015. Copy received from the author. November 11, 2015.

Koldanov, V. Ia. "Nekotorye itogi i vyvody po polezashchitnomu lesorazvedeniiu za istekshie piat' let." *Lesnoe khoziaistvo,* no. 3 (1954): 10–18.

Kolosov, Oleg. "Strakh peredaetsia russkim geneticheski rezul'tate bol'shevitskikh repressii?" December 30, 2013. http://rons-inform.livejournal.com/1303024 .html.

Kol'tsov, N. K. "Improvement of the Human Race." *Russkii evgenicheskii zhurnal,* no. 1 (1922): 3–27. Reprinted in *The Dawn of Human Genetics,* edited by V. V. Babkov, 66–86. Cold Spring Harbor, NY: Cold Spring Harbor Laboratory Press, 2013.

———. "Noveishie popytki dokazat' nasledstvennost' blagopreobretennykh priznakov." *Russkii evgenicheskii zhurnal,* no. 2 (1924): 159–167.

Kononkov, P. F. *Dva mira—dve ideologii. O polozhenii v biologicheskikh i sel'skokhoziaistvennykh naukakh v Rossii v sovetskii i postsovetskii period.* Moscow: Luch, 2014.

Korochkin, L. "Vo vlasti nevezhestva, Neolysenkovshchina v rossiiskom soznanii." *Literaturnaia gazeta,* March 6, 2002, 1–2.

Kouprianov, Alexei V. "The 'Soviet Creative Darwinism' (1930s–1950s): From the Selective Reading of Darwin's Works to the Transmutation of Species." *Studies in the History of Biology* 3, no. 2 (2011): 8–31.

Kouzarides, T. "Chromatin Modifications and Their Function." *Cell* 128 (February 23, 2007): 693–705.

Kozlov, Vladimir. "Strakh peredaetsia po nasledstvu?" *Gorodskie novosti,* July 25, 2013. http://www.gornovosti.ru/tema/neotlozhka/strakh-peredaetsya-po-nasledstvu40779.htm.

Krementsev, Nikolai. *International Science between the World Wars: The Case of Genetics.* London: Routledge, 2005.

———. "Marxism, Darwinism, and Genetics in the Soviet Union." In *Biology and Ideology: From Descartes to Dawkins,* edited by Denis R. Alexander and Ronald L. Numbers, 215–246. Chicago: University of Chicago Press, 2010.

———. *Stalinist Science.* Princeton, NJ: Princeton University Press, 1997.

Kundakovic, Marija, and Frances A. Champagne. "Early-Life Experience, Epigenetics, and the Developing Brain." *Neuropsychopharmacology,* July 30, 2014. doi: 10.1038/npp.2014.140.

Kuroedov, A. S. "Rol' sotsialisticheskoi sel'sko-khoziaistvennoi praktiki v razvitii michurinskoi biologii." Unpublished diss. for the degree of kandidat, Moscow State University, 1952.

Lachmann, M., and E. Jablonka. "The Inheritance of Phenotypes: An Adaptation to Fluctuating Environments." *Journal of Theoretical Biology* 181 (1996): 1–9.

Lamm, E., and E. Jablonka. "The Nurture of Nature: Hereditary Plasticity in Evolution." *Philosophical Psychology* 21, no. 3 (2008): 305–319.

Landman, Otto E. "The Inheritance of Acquired Characteristics." *Annual Review of Genetics* 21 (December 1991): 1–20.

Langdon-Davies, John. *Russia Puts the Clock Back: A Study of Soviet Science and Some British Scientists.* London: Gollancz, 1949.

Lartigue, Carole, John Glass, Nina Alperovich, Rembert Pieper, Prashanth Parmar, Clyde A. Hutchinson III, Hamilton O. Smith, and J. Craig Venter. "Genome Transplantation in Bacteria: Changing One Species to Another." *Science* 317 (August 3, 2007): 632–638.

Lecourt, Dominique. *Proletarian Science? The Case of Lysenko.* Atlantic Highlands, NJ: Humanities Press, 1977.

Lederberg, Joshua. "Cell Genetics and Hereditary Symbiosis." *Physiological Review* 32, no. 4 (1952): 403–430.

Lenz, F. "Der Fall Kammerer und seine Umfilmung durch Lunatscharsky." *Archiv für Rassen- und Gesellschafts Biologie,* no. 21 (1929): 311–318.

Lenz, Fritz. "Einführung in die Vererbungswissenschaft." *Archiv für Rassen- und Gesellschafts-Biologie* 21 (1929): 99–102.

Levit, S. G. "Evolutsionnye teorii v biologii i marksizm." *Meditsina i dialekticheskii materialism,* no. 1 (1926): 15–32.

Levites, E. V. "Epigenetic Variability as a Source of Biodiversity and a Factor of Evolution." Pt. 3. *Biodiversity and Dynamics of Ecosystems in North Eurasia* 1 (2000): 73–75.

Li, E., and A. Bird, "DNA Methylation in Animals." In *Epigenetics,* edited by C. D. Allis, T. Jenuwein, D. Reinberg, and M. L. Caparros, 341–356. Cold Spring Harbor, NY: Cold Spring Harbor Laboratory Press, 2004.

Lieberman, Daniel E. "Epigenetic Integration, Complexity, and Evolvability of the Head." In *Epigenetics: Linking Genotype and Phenotype in Development and Evolution,* edited by Benedikt Hallgrimsson and Brian K. Hall, 271–289. Berkeley: University of California Press, 2011.

Lindberg, Julia, Susanne Björnerfeldt, Peter Saetre, Kenth Svartberg, Birgitte Seehuus, Morten Bakken, Carlos Vila, and Elena Jazin. "Selection for Tameness Has Changed Brain Gene Expression in Silver Foxes." *Current Biology* 15, no. 22 (2005): R915–R916.

Liubishchev, A. A. *O monopolii T. D. Lysenko v biologii.* Ulyanovsk, Russia: Ul'ianovskii gosudarstvennyi pedagogicheskii universitet, 2004.

———. *O prirode nasledstvennykh faktorov.* Ulyanovsk, Russia: Ul'ianovskii gosudarstvennyi pedagogicheskii universitet, 2004.

Liubishchev, A. A., and A. G. Gurvich. *Dialog o biologii*. Ulyanovsk, Russia: Ul'ianovskii gosudarstvennyi pedagogicheskii universitet, 1998.

Logan, Cheryl A. *Hormones, Heredity, and Race: Spectacular Failure in Interwar Vienna*. New Brunswick, NJ: Rutgers University Press, 2013.

———. "Overheated Rats, Race, and the Double Gland: Paul Kammerer, Endocrinology and the Problem of Somatic Induction." *Journal of the History of Biology* 40, no. 4 (2007): 684–725.

Loukoianov, Artem, Liuling Yan, Ann Blechl, Alexandra Sanchez, and Jorge Dubcovsky. "Regulation of *VRN-1* Vernalization Genes in Normal and Transgenic Polyploid Wheat." *Plant Physiology* 138 (August 2005): 2364–2373.

Lumey, L. H. "Reproductive Outcomes in Women Prenatally Exposed to Undernutrition: A Review of Findings from the Dutch Famine Birth Cohort." *Proceedings of the Nutrition Society* 57, no. 1 (1998): 129–135.

Lumey, L. H., A. D. Stein, H. S. Kahn, K. M. van der Pal-de Bruin, G. J. Blauw, P. A. Zybert, and E. S. Susser. "Cohort Profile: The Dutch Hunger Winter Families Study." *International Journal of Epidemiology* 36, no. 6 (2007): 1194–1204.

Lunacharskii, Anatoli. *Lunacharskii o kino: Stat'i, vyskazyvaniia, stsenarii, dokumenty*. Moscow: Izdatel'stvo iskusstvo, 1965.

Lyon, Mary F. "Gene Action in the X-Chromosome of the Mouse." *Nature* 190, no. 4773 (1961): 372–373.

Lysenko, T. D. *Agrobiologiia*. Moscow: Sel'khozgiz, 1952.

———. "Gnezdovaia kul'tura lesa." *Ogonek*, no. 10 (March 1949): 6–7.

———. *Heredity and Its Variability*. Translated by Th. Dobzhansky. New York: King's Crown Press, 1946.

———, ed. *I. V. Michurin: Sochineniia v chetyrek tomakh*. Vols. 1–4. Moscow: Gosizdat, 1948.

———. "Iarovizatsiia—eto milliony pudov dobavochnogo urozhaia." *Izvestiia*, February 15, 1935, 4.

———. "Interesnye raboty po zhivotnovodstvu v Gorkakh Leninskikh." *Pravda*, July 17, 1957, 5–6.

———. *Izbrannye sochineniia*. Moscow: Moskovskii rabochii, 1953.

———. "Novoe v nauke o biologicheskom vide." *Pravda*, November 3, 1950, 2.

————. "Obnovlennye semena: Beseda s akademikom T. D. Lysenko." *Sotsialisticheskoe zemledelie,* September 16, 1935, 1.

————. "On vdokhnovlial nas na bor'bu za dal'neishii rastsvet nauki." *Izvestiia,* September 1, 1948, 1.

————. "Po povodu stat'i akademika N. I. Vavilova." *Sotsialisticheskoe zemledelie,* February 1, 1939, 25.

————. "Posev polezashchitnykh lesnykh polos gnezdovym sposobom." *Agrobiologiia,* no. 2 (1952). http://lysenkoism.narod.

————. "Rech' tovarishcha T. D. Lysenko." *Pravda,* February 26, 1956, 9.

————. "Shire primeniat' v nechernozemnoi polose organomineral'nye smesy." *Izvestiia,* April 27, 1957, 2.

————. "Teoreticheskie osnovy napravlennogo izmeneniia nasledstvennosti sel'skokhoziaistvennykh rastenii." *Pravda,* January 29, 1963, 3–4.

————. "Teoreticheskie uspekhi agronomicheskoi biologii." *Izvestiia,* December 8, 1957, 5.

————. "Vazhnye rezervy kolkhozov i sovkhozov." *Pravda,* March 14, 1959, 2–3.

————. "Vliianie termicheskogo faktora na prodolzhitel'nost' faz razvitiia rastenii." *Trudy azerbaidzhanskoi tsentral'noi opytnoselektsionnoi stantsii,* no. 3 (1928): 1–169.

Maderspacher, Florian. "Lysenko Rising." *Current Biology* 20, no. 19 (2010): R835–R837.

Maienschein, J. "Epigenesis and Preformationism." In *Stanford Encyclopedia of Philosophy,* edited by E. N. Zalta. Stanford, CA: Stanford University Press, 2008. http://plato.stanford.edu.

Maksimov, N. A. "Fiziologicheskie factory, opredeliaiuschie dlinu vegetatsionnogo perioda." *Trudy po prikladnoi botanike, genetike i selektsii* 20 (1929): 169–212.

Maksimov, N. A., and M. A. Krotkina. "Issledovaniia nad posledstviem ponizhennoi temperatury na dlinu vegetatsionnogo perioda." *Trudy po prikladnoi botanike, genetike i selektsii* 23, no. 2 (1929–1930): 427–473.

Manton, Kenneth G., and M. D. Golubovsky. "Three-Generation Approach in Biodemography Is Based on the Developmental Profiles and the Epigenetics of Female Gametes." *Frontiers in Bioscience* 10 (January 1, 2005): 187–191.

Martin, C., and Y. Zhang. "Mechanisms of Epigenetic Inheritance." *Current Opinion in Cell Biology* 19, no. 3 (2007): 266–272.

Mayr, E. *The Growth of Biological Thought: Diversity, Evolution, and Inheritance.* Cambridge, MA: Belknap Press of Harvard University Press, 1982.

McClintock, Barbara. "Introduction of Instability at Selected Loci in Maize." *Genetics* 38, no. 6 (1953): 579–599.

———. "The Origin and Behavior of Mutable Loci in Maize." *Proceedings of the National Academy of Sciences of the United States of America* 36, no. 6 (1950): 344–355.

———. "The Significance of Responses of the Genome to Challenge." *Science* 226, no. 4676 (1984): 792–801.

———. "Some Parallels between Gene Control Systems in Maize and in Bacteria." *American Naturalist* 95 (September–October 1961): 265–277.

McGowan, P. O., et al. "Epigenetic Regulation of the Glucocorticoid Receptor in Human Brain Associates with Child Abuse." *Nature Neuroscience* 12 (2009): 342–348.

Meaney, Michael J., Moshe Szyf, et al. "Epigenetic Programming by Maternal Behavior." *Nature Neuroscience* 7 (2004): 847–859.

Medvedev, Zhores. "Errors in the Reproduction of Nucleic Acids and Proteins and Their Biological Significance." Joint Publication Research Service, *Problems of Cybernetics,* no. 9 (November 1963): 21,448.

———. *The Rise and Fall of T. D. Lysenko.* New York: Columbia University Press, 1965.

Medvedev, Zhores, and V. Kirpichnikov. "Perspektivy sovetskoi genetiki." *Neva,* no. 3 (1963): 165–175.

Meloni, Maurizio. *Political Biology: Social Implications of Human Heredity from Eugenics to Epigenetics.* New York: Palgrave, 2016.

Merkel, Arkady L., and Lyudmila N. Trut. "Behavior, Stress, and Evolution in Light of the Novosibirsk Selection Experiment." In *Transformations of Lamarckism,* edited by Snait B. Gissis and Eva Jablonka, 171–180. Cambridge, MA: MIT Press, 2011.

Mikulinskii, S. R., ed. *Nikolai Ivanovich Vavilov: Ocherki, vospominaniia, materialy.* Moscow: Nauka, 1987.

Mironin, Sigizmund. *Delo genetikov*. Moscow: Algoritm, 2008.

Mitchell, P. Chalmers. "The Spencer-Weismann Controversy." *Nature* 49 (February 15, 1894): 373–374.

Molinier, J., G. Ries, C. Zipfel, and B. Hohn. "Transgeneration Memory of Stress in Plants." *Nature* 442 (2006): 1046–1049.

Morange, Michel. "L'épigénétique." *Études: Revue de culture contemporaine,* November 2014, 45–55.

Morgan, T. H. "Are Acquired Characters Inherited?" *Yale Review* (July 1924): 712–729.

Morgan, T. H., and Iu. A. Filipchenko. *Nasledstvennye li priobretennye priznaki?* Leningrad, USSR: Seiatel', 1925.

Morris, K. "Lamarck and the Missing Lnc." *Scientist,* October 1, 2012. http://www .the-scientist.com.

Mukhin, Iurii. *Prodazhnaia devka genetika*. Moscow: Izdatel'stvo Bystrov, 2006.

Muller, H. J. "Artificial Transmutation of the Gene." *Science* 66, no. 1699 (1927): 84–87.

———. "Lenin's Doctrines in Relation to Genetics." In *Science and Philosophy in the Soviet Union,* edited by Loren Graham, 453–469. New York: Alfred A Knopf, 1972.

———. "Letter from Muller to Stalin." In *The Dawn of Human Genetics,* edited by V. V. Babkov, 643–646. Cold Spring Harbor, NY: Cold Spring Harbor Laboratory Press, 2013.

———. "Nauka proshlogo i nastoiashchego i chem ona obiazana Marksu." In *Marksizm i estestvoznanie,* 204–207. Moscow: Izdatel'stvo kommunisticheskoi akademiii, 1933.

———. "Observations of Biological Science in Russia." *Scientific Monthly* 16, no. 5 (1923): 539–552.

———. *Out of the Night: A Biologist's View of the Future.* New York: Vanguard Press, 1935.

———. "Partial List of Biological Institutes and Biologists Doing Experimental Work in Russia at the Present Time." *Science* 57, no. 1477 (1923): 472–473.

Murneek, A. E., R. O. Whyte, et al., eds. *Vernalization and Photoperiodism: A Symposium.* Waltham, MA: Chronica Botanica, 1948.

Nanney, D. "Epigenetic Control Systems." *Proceedings of the National Academy of Sciences* 44 (1958): 712–717.

Noble, G. K. "Kammerer's *Alytes*." *Nature* 118 (August 7, 1926): 208–210.

Ol'shanskii, M. "Protiv fal'sifikatsii v biologicheskoi nauke." *Sel'skaia zhizn'* (August 1963): 2–3.

Ovchinnikov, N. V., P. F. Kononkov, A. Chichkin, and I. V. Driagena. *Trofim Denisovich Lysenko: Sovetskii agronom, biolog, selektsioner.* Moscow: Samoobrazovanie, 2008.

Painter, R. C., C. Osmond, et al. "Transgenerational Effects of Prenatal Exposure to the Dutch Famine on Neonatal Adiposity and Health in Later Life." *BJOG: Journal of Obstetrics and Gynaecology* 115, no. 10 (2008): 1243–1249.

Paul, Diane. *Controlling Human Heredity, 1865 to the Present.* Amherst, NY: Humanities Books, 1998.

———. "Eugenics and the Left." *Journal of the History of Ideas* 45, no. 4 (1984): 567–585.

Pembrey, Marcus E., Lars Olov Bygren, Gunnar Kaati, Sören Edvinsson, Kate Northstone, Michael Sjöström, Jean Golding, and the ALSPAC Study Team. "Sex-Specific, Male-Line Transgenerational Responses in Humans." *European Journal of Human Genetics* 14 (2006): 159–166.

Pennisi, Elizabeth. "The Case of the Midwife Toad: Fraud or Epigenetics?" *Science* 325 (September 4, 2009): 1194–1195.

Pigliucci, M. *Phenotypic Plasticity: Beyond Nature and Nurture.* Baltimore: Johns Hopkins University Press, 2001.

Platonov, G. "Dogmy starye i dogmy novye." *Oktiabr',* no. 8 (1965): 149–165.

Polianskii, V. I., and Iu. I. Polianskii. *Sovremennye problemy evoliutsionnoi teorii.* Leningrad, USSR: Nauka, 1967.

Popovsky, Mark. *The Vavilov Affair.* Hamden, CT: Archon Books, 1984.

Pringle, Peter. *The Murder of Nikolai Vavilov: The Story of Stalin's Persecution of One of the Great Scientists of the Twentieth Century.* New York: Simon and Schuster, 2008.

Provine, W. B. *Sewall Wright and Evolutionary Biology.* Cambridge, MA: MIT Press, 1986.

Purvis, O. N. "The Physiological Analysis of Vernalization." In *Encyclopedia of Plant Physiology,* edited by W. H. Ruhland, 16:76–117. Berlin: Springer, 1961.

Pyzhenkov, V. I. *Nikolai Ivanovich Vavilov—botanik, akademik, grazhdanin mira*. Moscow: Samoobrazovanie, 2009.

Rabkin, Yakov M. *Science between the Superpowers*. New York: Priority Press, 1988.

Rakyan, V. K., J. Preis, et al. "The Marks, Mechanisms and Memory of Epigenetic States in Mammals." Pt. 1. *Biochemical Journal* 356 (2001): 1–10.

Rando, O. J., and K. J. Verstrepen. "Timescales of Genetic and Epigenetic Inheritance." *Cell* 128, no. 4 (2007): 655–668.

Ratliff, Evan. "Taming the Wild." *National Geographic* (March 2011). http://ngm .nationalgeog-wild-animals/ratliff-text.

Reik, W. "Stability and Flexibility of Epigenetic Gene Regulation in Mammalian Development." *Nature* 447, no. 7143 (2007): 425–432.

Reik, W., W. Dean, et al. "Epigenetic Reprogramming in Mammalian Development." *Science* 293 (2001): 1089–1092.

Rheinberger, H.-J. "Gene." *Stanford Encyclopedia of Philosophy*. Stanford, CA: Stanford University Press, 2008. http://plato.stanford.edu.

Rheinberger, H.-J., and Staffan Müller-Wille. *A Cultural History of the Gene*. Chicago: University of Chicago Press, 2012.

Rice, William R., Urban Friberg, and Sergey Gavrilets. "Homosexuality as a Consequence of Epigenetically Canalized Sexual Development." *Quarterly Review of Biology* 87 (December 2012): 343–368.

Richards, E. J. "Inherited Epigenetic Variation—Revisiting Soft Inheritance." *Nature Reviews Genetics* 7, no. 5 (2006): 395–401.

Robertson, K., and A. Wolfe. "DNA Methylation in Health and Disease." *Nature Reviews Genetics* 1 (October 2000): 9–11.

Roll-Hansen, Nils. *The Lysenko Effect: The Politics of Science*. Amherst, NY: Humanity Books, 2005.

Roseboom, Tessa, Susanne de Rooij, and Rebecca Painter. "The Dutch Famine and Its Long-Term Consequences for Adult Health." *Early Human Development* 82 (2006): 485–491.

Rossianov, Kirill. "Editing Nature: Joseph Stalin and the 'New' Soviet Biology." *Isis* 84 (1993): 728–745.

———. "Stalin kak redaktor Lysenko." *Voprosy filosofii*, no. 2 (1993): 56–69.

Rukhkian, A. A. "Ob opisannom S. K. Karapetianom sluchae porozhdeniia lesh-chiny grabom." *Botanicheskii zhurnal* 38, no. 6 (1953): 885–891.

Rul'e, Karl Frantsevich. *Izbrannye biologicheskie proizvedeniia.* Moscow: Izdatel'stvo AN SSSR, 1954.

Santos, F., and W. Dean. "Epigenetic Reprogramming during Early Development in Mammals." *Reproduction* 127, no. 6 (2004): 643–651.

Sapp, Jan. *Beyond the Gene: Cytoplasmic Inheritance and the Struggle for Authority in Genetics.* New York: Oxford University Press, 1987.

Schmalhausen, I. I. *Factors of Evolution: The Theory of Stabilizing Selection.* Philadelphia: Blackiston, 1949.

Schubeler, D. "Epigenomics: Methylation Matters." *Nature* 462, no. 7271 (2009): 296–297.

Schuster, Gernot. "Paul Kammerer: Der Krötenküsser." In *Haltestelle Puchberg am Schneeberg: Porträts berühmter Gäste und Gönner,* edited by Gernot Schuster and · Peter Zöchbauer, 202–218. Puchberg, Austria: Verlag Berger, n.d.

Schwartz, James. *In Pursuit of the Gene: From Darwin to DNA.* Cambridge, MA: Harvard University Press, 2008.

Serebrovsky, A. S. "Anthropogenetics and Eugenics in a Socialist Society." In *The Dawn of Human Genetics,* edited by V. V. Babkov, 505–518. Cold Spring Harbor, NY: Cold Spring Harbor Laboratory Press, 2013.

———. "Teoriia nasledstvennosti Morgana i Mendelia i marksisty." *Pod znamenem marksizma,* no. 3 (1926): 98–117.

Seyla, Rena. "Defending Scientific Freedom and Democracy: The Genetics Society of America's Response to Lysenko." *Journal of the History of Biology* 45, no. 1 (2012): 415–442.

Shatskii, A. L. "K voprosu o summe temperatur, kak sel'skokhoziaistvenno-klimaticheskom indekse." *Trudy po sel'skokhoziaistvennoi meteorologii* 21, no. 6 (1930): 261–263.

Sinnott, E. W., L. C. Dunn, and Th. Dobzhansky. *Principles of Genetics.* New York: McGraw-Hill, 1950.

Situation in Biological Science: Proceedings of the Lenin Academy of Agricultural Sciences of the U.S.S.R., July 31–August 7, 1948. Complete stenographic report, New York, 1949. Also in Russian: *O polozhenii v biologicheskoi nauke.* Stenograficheskii otchet sessii vsesoiuznoi akademii sel'sko-khoziaistvennykh nauk imeni V. I. Lenina, 31 iiulia–7 avgusta 1948 g., Moscow, 1948.

Skipper, Magdalena, Alex Eccleston, Noah Gray, Therese Heemels, Nathalie Le Bot, Barbara Marte, and Ursula Weiss. "Presenting the Epigenome Roadmap." *Nature* 518, no. 313 (February 19, 2015). doi: 10.1038/518313a.

Slepkov, Vasilii. "Nasledstvennost' i otbor u cheloveka (Po povodu teoreticheskikh predposylok evgeniki)." *Pod znamenem marksizma,* no. 4 (1925): 102–122.

Slotkin, R., and R. Martienssen. "Transposable Elements and the Epigenetic Regulation of the Genome." *Nature Reviews Genetics* 8, no. 4 (2007): 272–285.

Smirnov, E. S. "Novye dannye o nasledstvennom vliianii sredy i sovremennyi lamarkizm." *Vestnik kommunisticheskoi akademii* 25 (1928): 197.

———. *Problema nasledovaniia probretennykh priznakov: Kriticheskii obzor literatury.* Moscow: Izdatel'stvo Komakademii, 1927.

Smirnov, E. S., Iu. M. Vermel', and B. S. Kuzin. *Ocherki po teorii evoliutsii.* Moscow: Krasnaia Nov', 1924.

Smith, Martin. "How Good Is Soviet Science?" *NOVA,* WGBH, 1987. Loren Graham, rapporteur.

Sokolov, B. "Ob organizatsii proizvodstva gibridnykh semian kukuruzy." *Izvestiia,* February 2, 1956, 2.

Sokolov, V. A. "Imprinting in Plants." *Russian Journal of Genetics* 42, no. 9 (2006): 1043–1052.

Sonneborn, T. M. "H. J. Muller, Crusader for Human Betterment." *Science* (November 15, 1968): 772.

———. "Heredity, Environment, and Politics." *Science* 111 (1950): 535.

Soyfer, Valery. *Lysenko and the Tragedy of Soviet Science.* New Brunswick, NJ: Rutgers University Press, 1994.

Spector, T. D. *Identically Different: Why You Can Change Your Genes.* London: Weidenfeld and Nicolson, 2012.

Stebbins, G. L. *Darwin to DNA, Molecules to Humanity.* San Francisco: Freeman, 1982.

Stegemann, Sandra, and Ralph Bock. "Exchange of Genetic Material between Cells in Plant Tissue Grafts." *Science* 324 (May 1, 2009): 649–651.

Stolz, K., P. E. Griffiths, et al. "How Biologists Conceptualize Genes: An Empirical Study." *Studies in the History and Philosophy of Biological and Biomedical Sciences* 35 (2004): 647–673.

Sturtevant, A. H. *A History of Genetics*. Cold Spring Harbor, NY: Cold Spring Harbor Laboratory Press, 2001.

Sukhachev, V. N. "O vnutrividovykh i mezhvidovykh vzaimootnosheniiakh sredi rastenii." *Botanicheskii zhurnal* 38, no. 1 (1953): 57–96.

Sukhachev, V. N., and N. D. Ivanov. "K voprosam vzaimootnoshenii organizmov i teorii estestvennogo otbora." *Zhurnal obshchei biologii* 15, no. 4 (July–August 1954): 303–319.

Surani, M. A. H., S. C. Barton, and M. L. Norris. "Development of Reconstituted Mouse Eggs Suggests Imprinting of the Genome during Gametogenesis." *Nature* 308 (April 5, 1984): 548–550.

Susiarjo, Martha, and Marisa S. Bartolomei. "You Are What You Eat, But What about Your DNA? Parental Nutrition Influences the Health of Subsequent Generations through Epigenetic Changes in Germ Cells." *Science* 345 (August 15, 2014): 9–10.

Szyf, M., et al. "Maternal Programming of Steroid Receptor Expression and Phenotype through DNA Methylation in the Rat." *Frontiers in Neuroendocrinology* 26, nos. 3–4 (October–December 2005): 139–162.

Takahashi, K., and S. Yamanaka. "Induction of Pluripotent Stem Cells from Embryonic and Adult Fibroblast Cultures by Defined Factors." *Cell* 126 (August 26, 2006): 663–676.

Targul'ian, O. M., ed. *Spornye voprosy genetiki is selektsii: Raboty IV sessii akademii 19–27 dekabriia 1936 goda*. Moscow and Leningrad: VASKHNIL, 1937.

Taschwer, Klaus. *A Movie That Shaped the History of Biology (a Bit)—On the Soviet-German Silent Movie "Salamandra" from 1928*. Report given at the Max-Planck-Institut für Wissenschaftsgeschichte, June 20, 2012.

———. "The Toad Kisser and the Bear's Lair: The Case of Paul Kammerer's Midwife Toad Revisited." Report given at the Max-Planck-Institut für Wissenschaftsgeschichte, October 8, 2015, 4.2011–6.2012.

Timofeev-Resovskii, N. V. *Istorii, rasskazannye im samim, s pis'mam, fotografiiami i dokumentami*. Moscow: Soglasie, 1998.

Tobi, E. W., L. H. Lumey, et al. "DNA Methylation Differences after Exposure to Prenatal Famine are Common and Timing- and Sex-Specific." *Human Molecular Genetics* 18, no. 21 (2009): 4046–4053.

Todes, Daniel. *Ivan Pavlov: A Russian Life in Science*. New York: Oxford University Press, 2014.

Trut, L. N., I. Z. Plyusnina, and I. N. Oskina. "An Experiment on Fox Domestication and Debatable Issues of Evolution of the Dog." *Russian Journal of Genetics* 40, no. 6 (2004): 644–655.

Urnov, F. D., and A. P. Wolffe. "Above and Within the Genome: Epigenetics Past and Present." *Journal of Mammary Gland Biology and Neoplasia* 6, no. 2 (2001): 153–167.

Vanyushin, B. F. "DNA Methylation and Epigenetics." *Russian Journal of Genetics* 42, no. 9 (2006): 985–997.

Verdin, Eric, and Melanie Ott. "50 Years of Protein Acetylation: From Gene Regulation to Epigenetics, Metabolism, and Beyond." *Nature Reviews Molecular Cell Biology*. doi: 10.1038/nrm3931.

Vert'yanov, S. Iu. "Great Damage." *Obshchaia Biologiia* (2012): 198.

Vladimirskii, A. P. *Peredaiutsia li po nasledstvu priobretennye priznaki?* Moscow-Leningrad: Izdatel'stvo komakademii, 1927.

Vöhringer, Margarete. "Behavioural Research, the Museum Darwinianum and Evolutionism in Early Soviet Russia." *History and Philosophy of the Life Sciences* 31 (2009): 279–294.

Volotskoi, M. V. *Fizicheskaia kul'tura s tochki zreniia evgeniki*. Moscow: Izdatel'stvo instituta fizicheskoi kul'tury imeni P. F. Lesgafta, 1924.

———. *Klassovye interesy i sovremennaia evgenika*. Moscow: Zhizn' i znanie, 1925.

Vorontsov, N. "Zhizn' toropit: Nuzhny sovremennye posobiia po biologii." *Komsomol'skaia Pravda*, November 11, 1964, 2.

Vucinich, Alexander. *Empire of Knowledge: The Academy of Sciences of the USSR, 1917–1970*. Berkeley: University of California Press, 1984.

Waddington, C. H. *The Evolution of an Evolutionist*. Edinburgh: Edinburgh University Press, 1975.

———. *New Patterns in Genetics and Development*. New York: Columbia University Press, 1968.

Wagner, Günther. "Paul Kammerer's Midwife Toads: About the Reliability of Experiments and Our Ability to Make Sense of Them." *Journal of Experimental Zoology* 312B (2009): 665–666.

Wang, Jessica. *American Scientists in an Age of Anxiety: Scientists, Anticommunism, and the Cold War.* Chapel Hill: University of North Carolina Press, 1999.

Weaver, I. C. G., N. Cervoni, F. A. Champagne, A. C. D'Alessio, S. Sharma, J. R. Seckl, S. Dymov, M. Szyf, and M. J. Meaney. "Epigenetic Programming by Maternal Behavior." *Nature Neuroscience* 7, no. 8 (August 2004): 847–854.

Weiner, Douglas. *A Little Corner of Freedom: Russian Nature Protection from Stalin to Gorbachev.* Berkeley: University of California Press, 1999.

———. "The Roots of 'Michurinism': Transformist Biology and Acclimatization as Currents in the Russian Life Sciences." *Annals of Science* 42, no. 3 (1985): 243–260.

Weissmann, Gerald. "The Midwife Toad and Alma Mahler: Epigenetics or a Matter of Deception?" *FASEB Journal* 24 (August 2010): 2591–2595.

Whitelaw, Emma. "Epigenetics: Sins of the Fathers, and Their Fathers." *European Journal of Human Genetics* 14 (2006): 131–132.

Winther, R. G. "August Weismann on Germ-Plasm Variation." *Journal of the History of Biology* 34 (2001): 517–555.

Wolfe, Audra Jayne. "What It Means to Go Public: The American Response to Lysenkoism, Reconsidered." *Historical Studies in the Natural Sciences* 40 (2010): 48–78.

Wright, Sewall. "Dogma or Opportunism?" *Bulletin of the Atomic Scientists* (May 1949): 141–142.

Wu, C. T., and J. R. Morris. "Genes, Genetics, and Epigenetics: A Correspondence." *Science* 293, no. 5532 (2001): 1103–1105.

Yool, A., and W. J. Edmunds. "Epigenetic Inheritance and Prions." *Journal of Evolutionary Biology* 11, no. 2 (1998): 241–242.

Young, Emma. "Rewriting Darwin: The New Non-Genetic Inheritance." *New Scientist* (July 9, 2008): 28–33.

Youngblood, Denise J. "Entertainment or Enlightenment? Popular Cinema in Soviet Society, 1921–1931." In *New Directions in Soviet History,* edited by Stephen White, 41–50. Cambridge: Cambridge University Press, 2002.

Youngson, N. A., and E. Whitelaw. "Transgenerational Epigenetic Effects." *Annual Review of Genomics and Human Genetics* 9, no. 1 (2008): 233–257.

Zavadovskii, B. M. *Darvinizm i marksizm*. Moscow: Gosudarstvennoe izdatel'stvo, 1926.

Zhivotovskii, L. A. "A Model of the Early Evolution of Soma-to-Germline Feedback." *Journal of Theoretical Biology* 216, no. 1 (2002): 51–57.

Zhivotovskii, Lev. *Neizvestnyi Lysenko*. Moscow: T-vo nauchnykh izdanii KMK, 2014.

Zimmermann, W. *Vererbung erzwungener Eigenschaften und Auslese*. Jena, Germany: Verlag von Gustav Fischer, 1938.

Zirkle, Conway. "The Early History of the Idea of the Inheritance of Acquired Characters and of Pangenesis." *Transactions of the American Philosophical Society*, n.s., 35, pt. 2 (1946): 91–151.

———. "Further Notes on Pangenesis and the Inheritance of Acquired Characteristics." *American Naturalist* 70, no. 731 (1936): 529–546.

———. "The Inheritance of Acquired Characteristics and the Provisional Hypothesis of Pangenesis." *American Naturalist* 69, no. 724 (1935): 417–415.

ACKNOWLEDGMENTS

First, I would like to thank the American and Russian institutions and foundations that have supported my research on Russian science over the past several decades. They include the U.S. government through programs such as international scholarly exchanges between the United States and the Soviet Union and Russia, and Russian institutions such as the Russian Academy of Sciences, the European University in St. Petersburg, Moscow University, and many other Russian universities, institutes, libraries, and archives. On the American side, CRDF Global, the National Academy of Sciences, the Fulbright-Hayes Program, the International Research and Exchanges Program, the Guggenheim Foundation, the Institute for Advanced Study, the Ford Foundation, the John D. and Catherine T. MacArthur Foundation, the Carnegie Corporation of New York, the Social Science Research Council, the National Science Foundation, the National Humanities Foundation, and the Sloan Foundation have all, at one time or another, supported my work during dozens of trips to Russia.

The universities where I have been on the faculty—Indiana University, Columbia University, the Massachusetts Institute of Technology, and Harvard University—have been generous in their assistance both to my teaching and to my research on Russian science. I have been fortunate to be a citizen of a country in which there exists such a variety of organizations supporting scholarship, both private and governmental. And although much more financial support for my research has come from American sources than from Russian ones, I am also indebted to Russia for its hospitality, at both favorable and unfavorable moments politically, which allowed me to work for many months, even several years, in Russian libraries and archives. As briefly mentioned in this book, Russian government officials showed their disapproval of my work by declaring me persona non grata, but several years later, the Russian Academy of Sciences awarded me a medal at a ceremony in Moscow for "distinguished work in the history of science." The scholarly work of an American on Russia can be a little complicated and even difficult, but on the whole I have been impressed with the willingness of the Russians to allow me to research their records; sometimes they have even said complimentary things about the results. I would like to think that if a Russian scholar spent as much time in the United States working on American science as I have done in Russia working on Russian science, the reaction of Americans would be at least as complimentary, but I will never know.

The pioneers in work on the history of Lysenkoism were David Joravsky, Zhores Medvedev, and Valery Soyfer, to whom I am deeply grateful. Other important contributors, most of them a bit later, were Mark Adams, A. E. Gaissinovitch, Nikolai Krementsev, Alexei B. Kojevnikov, Nils Roll-Hansen, Eduard Kolchinsky, Nikolai Vorontsov, Mikhail Golubovsky, and Mikhail Konashev. Many of their works are listed in the bibliography. My fellow historians of science Eduard Kolchinskii and Mikhail Konashev shared with me their research on biology in Russia and the Soviet Union. In their home city, St. Petersburg, I became friends with Sergei Inge-Vechtomov, head of genetics at St. Petersburg University and a person deeply knowledgeable about the history of his field. In the same city, I interviewed V. A. Baranov of the Russian Academy of Medical Sciences and Lidiia Khoroshinina, who has done very valuable work on the effects of the Leningrad Blockade during World War II on the health of the survivors. In Moscow I interviewed many biologists and talked with Trofim Lysenko.

Assistance on researching Paul Kammerer in Austria came especially from the director of the Puchberg Museum, Dr. Karl Rieter; Manuela Hödl of the Puchberg information office; and Sonja Marg. Bauer. It was striking to me that an unknown American, after walking into a little village in Austria and asking questions about events almost ninety years earlier, could find such encouragement and help in locating local sources. Sonya Bauer and her husband invited my wife and me into their home, where she produced local materials on the fate of Paul Kammerer in her hometown.

In Russia I met similar hospitality. Nikolai Vorontsov, a distinguished biologist and parliamentarian, told me stories of his participation in the long struggle against Lysenko. Nikolai served as the first and last minister of the environment in the Soviet Union; he was the only non-Communist ever to be a member of the Council of Ministers of the Soviet Union, the equivalent in the United States of the president's cabinet. Nikolai and his entire family became close friends. His wife, Elena Lyapunova, a biologist at the Koltsov Institute of Developmental Biology of the Russian Academy of Sciences, also knew much about the Lysenko affair and the history of biology in Russia. Her institute is named after the Nikolai Kol'tsov featured in Chapter 4 in this book. Nikolai and Elena's daughter Masha and son George are also close friends who have assisted me in numerous ways.

In Novosibirsk, Dmitrii Belyaev hosted me several times at his famous experiment station, where, for the first time, he successfully domesticated wild foxes. In his initial conversations with me, Belyaev was strongly opposed to any idea of the inheritance of acquired characteristics; however, as he worked with his fox subjects he began to think that "stress" had genetic effects.

The person who first gave me the idea of working specifically on epigenetics in Russia was Thomas Byrne, professor of neurology at Harvard Medical School. In a vivid conversation in his home, he wondered how epigenetics was being accepted in Russia. I soon investigated, and a different world of scholarship opened up. Eva Jablonka told me about a conference that she and her colleague Marion J. Lamb attended there and how epigenetics became a heated topic of conversation. Maurizio Meloni, who is writing a book on political biology that intersects with this one in many ways, engaged me with sparkling conversation that helped me in the understanding of many issues.

Friends and fellow scholars who have read different versions of the manuscript and given me valuable advice include David Tatel, Michael Gordin, Vyacheslav Gerovitch, Gerson Sher, Marjorie Senechal, Felton James (Tony) Earls, Mary Carlson, Edward O. Wilson, Jerome Kagan, Michael Golubovsky, Michael Meaney, and Douglas Weiner. Each more than deserves further mention.

David Tatel is a distinguished judge, but even more he is an intellectual with a voracious appetite. He reads widely; enough to read an early version of this manuscript and to make many useful comments. Michael Gordin is leading a group of scholars at Princeton University studying the history of science in Russia and making a major impact on the field, both through his own scholarship and through that of his students and former students. I am fortunate to benefit from his advice. Slava Gerovitch has written incisive works on cybernetics, the Soviet space program, and mathematics. I am the beneficiary of his comments on my own work. Gerson Sher is both a talented administrator and a scholar—a person who made an enormous impact in my field by founding the Civilian Research and Development Foundation (CRDF). He is generous in his assistance to others, including me. Marjorie Senechal is a mathematician and a universal scholar who has energetically promoted cooperative scientific work between American and Russian scientists. She has been unstinting in working with me, helping my own efforts to improve. I traveled numerous times in Russia with Gerson Sher, Marjorie Senechal, Victor Rabinowitch, Marilyn Pifer, and Harley Balzer, all of whom enriched my knowledge. Tony Earls and his wife, Mary Carlson, are more knowledgeable on many subjects than I and have helped me come closer to understanding some very difficult issues in epigenetics. Edward O. Wilson is one of the leading biologists of his generation. He read the manuscript and helped me find an agent and, ultimately, an outstanding publisher. Jerry Kagan is omnivorous in his interests while pursuing his own distinguished scholarship. He has generously read my work and commented on it. Michael Golubovsky is that rare combination of research scientist and historian, and his knowledge of the recent history of genetics and molecular biology is deep. He made many suggestions to me for further reading and commented critically upon my writings. Michael Meaney is one of the best-known researchers on transgenerational epigenetic inheritance in the world. I have benefited from his comments on the book manuscript. Douglas Weiner is the leading historian of environmental issues in Russia and the Soviet Union and in that capacity encounters many questions close to the subject of this book, such as Lysenkoism and the complicated history of Soviet biology. He also read an early version of the book and contributed to my better understanding of my own work. The distinguished geneticist Vicki Chandler inspired me with a comment at a meeting of the American Philosophical Society in Philadelphia.

There is no better place than Cambridge, Massachusetts, to try out ideas on friends and colleagues; among those who have helped me the most are Howard Gardner, Ellen Winner, Shelly Greenfield, Allan Brandt, Kay Merseth, and Charles Rosenberg. Two of my dearest friends are Everett Mendelsohn and Mary Anderson, both of whom have enlightened me on a great diversity of topics, including the "accursed questions" of life. Everett is a distinguished historian of biology who has trained a generation of people in the field. Mary has deep knowledge of international development issues and many other subjects as well. It has been a personal joy to share their friendship.

I have done much of the writing of this book at the Davis Center for Russian and Eurasian Studies at Harvard University, which has generously provided me with an office and assistance. The directors of the center, Terry Martin and Rawi Abdelal, and the executive director, Alexandra Vacroux, created an atmosphere perfect for scholars, and several of the members of the staff have helped me with specific tasks. Rebekah Judson rescued a bibliography that had disappeared into a mysterious electronic cloud, Donna Griesenbeck helped me find research assistance, Sarah Failla was always present with appropriate advice, Katie Genovese and Maria Altamore took care of dozens of office tasks, and Penelope Skalnik arranged several speaking engagements. Hugh Truslow, the librarian for the Davis Center, is an expert on finding elusive information.

My agent, Ike Williams, and his coworker Katherine Flynn have expertly guided this book to a wonderful publisher. It was Ike who suggested the title for the book. The Harvard University Press is a true friend of scholars, and I have benefited from the guidance of its staff, especially that of my editor, Kathleen McDermott.

My wife, Patricia Albjerg Graham, has been my companion for over sixty years. In all that time, we have never run out of things to talk about. A fellow scholar, she has saved me from many errors (unfortunately, not quite all). With our daughter, Meg, and her husband, Kurt, we enjoy life fully. This book is dedicated to Meg and Kurt. I am fortunate to have such a family.

INDEX